Condition-Based Maintenance and Residual Life Prediction

Scrivener Publishing
100 Cummings Center, Suite 541J
Beverly, MA 01915-6106

Publishers at Scrivener
Martin Scrivener (martin@scrivenerpublishing.com)
Phillip Carmical (pcarmical@scrivenerpublishing.com)

Condition-Based Maintenance and Residual Life Prediction

Edited by

Chandan Deep Singh

Davinder Singh

Kanwal Jit Singh

and

Harleen Kaur

Scrivener Publishing

WILEY

This edition first published 2025 by John Wiley & Sons, Inc., 111 River Street, Hoboken, NJ 07030, USA and Scrivener Publishing LLC, 100 Cummings Center, Suite 541J, Beverly, MA 01915, USA

For more information about Scrivener publications please visit www.scrivenerpublishing.com.

Wiley Global Headquarters
111 River Street, Hoboken, NJ 07030, USA

For details of our global editorial offices, customer services, and more information about Wiley products visit us at www.wiley.com.

Library of Congress Cataloging-in-Publication Data

ISBN 978-1-119-93312-0

Front cover images supplied by Adobe Firefly
Cover design by Russell Richardson

Set in size of 11pt and Minion Pro by Manila Typesetting Company, Makati, Philippines

Printed in the USA

10 9 8 7 6 5 4 3 2 1

Contents

1

Maintenance

Harpreet Sharma[1]*, Chandan Deep Singh[2] and Kanwal Jit Singh[3]

[1]*Yadavindra Department of Engineering, Guru Kashi Campus, Talwandi Sabo, Punjab, India*
[2]*Department of Mechanical Engineering, Punjabi University, Patiala, India*
[3]*School of Agriculture Sciences and Engineering, Maharaja Ranjit Singh Punjab Technical University, Bathinda, India*

Abstract
In order to stay competitive, it is necessary for companies to continuously increase the effectiveness and efficiency of their production processes. Production strategies such as Justin-Time and Lean Production demand high availability of production equipment in order to meet customer satisfaction. Maintenance is a critical aspect of managing assets and ensuring their optimal performance and longevity. It involves a range of activities aimed at preserving, repairing, and optimizing equipment, systems, and infrastructure to minimize downtime, increase reliability, and reduce costs. This chapter provides an overview of maintenance practices, including preventive, corrective, and predictive maintenance, as well as their benefits and challenges.

***Keywords*:** Preventive maintenance, corrective maintenance, routine maintenance, scheduled maintenance, predictive maintenance, breakdown maintenance

1.1 Introduction and Meaning

Past and current maintenance practices in both private and government sectors would imply that maintenance is the action associated with equipment repair after it is broken. The dictionary defines maintenance as "the work of keeping something in proper condition, upkeep." This would imply that maintenance should be actions taken to prevent a device or component from failing or to repair normal equipment degradation experienced

**Corresponding author*: harpreetsharma3211@gmail.com

Chandan Deep Singh, Davinder Singh, Kanwal Jit Singh and Harleen Kaur (eds.) Condition-Based Maintenance and Residual Life Prediction, (1–22) © 2025 Scrivener Publishing LLC

with the operation of the device to keep it in proper working order. Data obtained in many studies over the past decade indicates that most private and government facilities do not expend the necessary resources to maintain equipment in proper working order. They wait for equipment failure to occur and then take whatever actions are necessary to repair or replace the equipment. Nothing lasts forever and all equipment has a predefined life expectancy or operational life associated with it.

1.2 Need for Maintenance

The need for maintenance is based on the actual or imminent failure. Ideally, maintenance is carried out to keep equipment and systems running efficiently during at least its usual life cycle (Lai *et al.*, 2019). As such, the practical functioning of equipment is based on time. If you want to graphically represent the failure rate of a piece of equipment in relation to time, it is probable that the graphic will take the shape of a bathtub, such as the one shown in Figure 1.1, where the Y-axis represents the failure rate and the X-axis represents time. This curve can be divided into three periods: premature death, lifecycle, and exhaustion period.

The first period, the premature death, of the curve is characterized by a high rate of failure, followed by a decreasing failure period. Most of the failures related at this time are linked to bad design, bad installation, or wrong use. The premature death period is followed by a period with a nearly constant failure rate known as the lifecycle. There are many theories

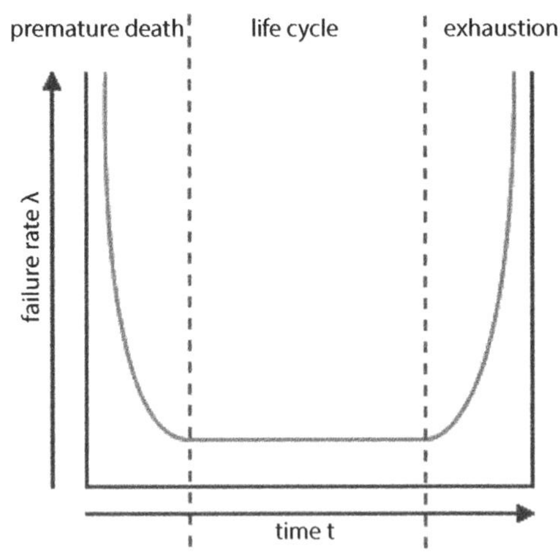

Figure 1.1 Product failure rate.

about why equipment fails in this time area. Most of them agree that poor preventive maintenance has often a key role. It is also generally agreed that exceptional practices related to predictive and preventive maintenance can extend this period. The exhaustion period is characterized by an increasing rate of failure. In most cases, this period includes a regular distribution of failures during the design life but in reverse (Jian *et al.*, 2021).

The lifecycle of most equipment needs periodic maintenance. If we use the example of a car, we could say that filters must be changed, front-end alignment must be maintained, oil change and proper lubrication are needed, and so on. In some cases, certain pieces need to be replaced, for example, the timing belt, to ensure the proper functioning of the main piece of the equipment, in this particular case, the car, once its design life has ended. Each time we do not carry out the maintenance activities planned by the designer, we shorten the operational lifecycle of the equipment.

1.3 Importance of Maintenance

- *Operator and Machine Safety*

By performing regular preventive maintenance, you will always be assured that your equipment will operate under safe conditions, both for the machine and the operators. Possible issues can be cut off before they cause any harm, without taking any chance.

- *Machine Efficiency*

With normal wear and tear, machines can cause lower efficiency. Proper preventive maintenance management will assure you of the optimal working conditions of the equipment and, moreover, conserve its life span.

- *Time Savings*

Planned preventive maintenance management will reduce the actual downtime caused by the breakdown and further enhance the products in less time, thereby saving a lot of time. Preventive maintenance will consume less time than emergency repairs and replacements.

1.4 Objectives of Maintenance

Equipment is an important resource that is constantly used for adding value to products. So, it must be kept in the best operating condition. Otherwise, there will be excessive downtime and also interruption of production if it is used in a mass production line. Poor working conditions of equipment will lead to quality-related problems. Hence, it is an absolute necessity to maintain equipment in good operating conditions at an economical cost. Hence, we need an integrated approach to minimize the cost of maintenance. In certain cases, the equipment will be obsolete over a period of time. If a firm wants to be in the same business competitively, it has to decide on whether to replace the equipment or retain the old equipment by taking the cost of maintenance and operation into account (Stone *et al.*, 2020).

1.5 The Role of the Maintenance Department

Responsibilities of a Maintenance Department

1. Engineering and execution of planned maintenance: The responsibility for the continued economical use of the equipment lies with the operating supervisor. It is, however, the responsibility of the maintenance staff, in cooperation with the operating supervisors to organize preventive maintenance schedules and improve the equipment efficiency necessary to meet the production schedule requirements.
2. Generation and distribution of power and other utilities: The power engineer is responsible for the generation and distribution of utilities like steam and compressed air in suitable quantities, as well as the purchase and distribution of electrical power, water, etc., at minimum cost.
3. Stand-by power generation: Apart from purchasing electric power from electricity board (EB), if a stand-by diesel generator is available in the factory, the maintenance staff would be responsible for operating this generator and supplying sufficient power during power outages.
4. Administrative and staffing: To handle mechanical work involved in carrying out the functions of the maintenance department, an adequate skilled labor force, provided with suitable equipment and proper supervision, is necessary. The type and size of this group are determined based on a balance between the costs of establishing and maintaining the group versus the cost of contracting for the maintenance work to outside firms, considering the factor of emergency and round-the-clock service. It is the responsibility of the head of the maintenance department to plan and coordinate work

distribution among the crafts and to organize and carry out programs for the training of the supervisors and craftsmen.

5. Engineering and supervision of construction projects within the scope of the group: Maintenance engineering is a definite part of the preliminary and final design of construction projects. It is the responsibility of the maintenance group to be familiar with all the projects under consideration or execution that will eventually become a part of the plant.
6. Technical consultation: The maintenance engineer would act as a consultant to the production department in solving the mechanical problem of the equipment.
7. Plant safety activities and equipment: The provision and control of all safety equipment, their maintenance, and monitoring for the general observance of plant safety practices by all workforces in the plant, would be the responsibility of the maintenance department.
8. Plant insurance: Strategies for securing plant insurance and ensuring the preservation of all critical data and documents, as well as taking the necessary action in the event of accidents.
9. Service facilities: Engineering, administration, and maintenance of various service facilities delegated to the departments like yards, roadways, washing rooms, sewers and waste disposal sections, and employee welfare services within the budgets provided by the management.
10. Maintenance of plant and equipment records: Maintaining the property and all relevant records of all plant and equipment required for the annual accounts and audit.
11. Meticulous breakdown record-keeping for future maintenance planning: Keeping the breakdown record for each and every piece of machinery and equipment meticulously as required for future maintenance planning.
12. Efficient execution of all functions: Last but not least, performing all the above functions in a safe and efficient manner.

1.6 Responsibilities of a Maintenance Engineer

Reliabilityweb.com suggests the following functions for the maintenance engineer.

1. Ensures that the equipment is properly designed, selected, and installed based on the lifecycle philosophy.

2. Ensures that the equipment is performing effectively and efficiently.
3. Establishes and monitors programs for engine and compressor analysis, as well as vibration and other condition monitoring techniques.
4. Reviews deficiencies noted during corrective maintenance.
5. Provides technical guidance for the Computerized Maintenance Management System (CMMS).
6. Maintains and advises on the use and disposition of stock, surplus, and rental equipment.
7. Promotes equipment standardization, recommends spare part levels, and coordinates sharing of spare parts with other asset teams.
8. Becomes available for consultation with maintenance technicians.
9. Monitors new technology and keeps management and staff apprised of the new developments.
10. Champions quality assurance services including shop qualifications for outside services.
11. Develops standards and procedures for major maintenance jobs.
12. Periodically makes cost/benefit reviews of maintenance management programs for areas of responsibility and exchanges information across asset reams.
13. Providing technical guidance for preventive (PM) and predictive (PDM) maintenance programs.
14. Monitors competitors' activities in the field of maintenance management.
15. Becomes the focal point for monitoring performance indicators for maintenance management programs.
16. Optimizes maintenance strategies.
17. Becomes a focal point for analyzing equipment operating data, root cause analysis, and ultrasonic and vibration analysis.
18. Imparts maintenance training.

1.7 Principles of Maintenance

They are followed in a system to guide the staff to work efficiently and effectively to achieve the overall objectives of the maintenance system.

Plant management in maintenance work:
The main role of a maintenance function is to provide safe and effective operation of the equipment to achieve the desired targets on time with economic usage of resources.

Production and maintenance objectives:
The plant operation is driven by the production targets. The objective of the maintenance function is to support these targets. The achievement of the desired goals of the production system is to be supported by both the production and maintenance departments to ensure the smooth and successful operation of the industry.

Establishment of work order and recording system:
The maintenance system should have a proper work and recording system. The work order for the maintenance function indicates the nature of the work to be performed and the series of operations to be followed to execute a particular job. It is necessary to maintain proper records and entries to monitor the maintenance function.

Information-based decision-making:
The maintenance objectives are successfully achieved by the use of a reliable information system. This information is used to meet the manpower and spare parts requirements of the industry.

Adherence to planned maintenance strategy:
Sound maintenance management should adhere to the planned maintenance strategy. This also includes the use of manufacturer information on the life and maintenance schedules of the equipment and other material resources available.

Planning of maintenance function:
All the maintenance functions are to be carefully executed through proper planning to ensure the effective utilization of manpower and materials.

Manpower for maintenance:
The manpower requirement of the maintenance system must be carefully evaluated based on the time and motion study. 'Ihe requirements should also satisfy the need arising in case of overhauls, component replacement, emergency, and unscheduled repair.

Workforce control:
Determining the exact workforce required to meet the maintenance objectives of the system is a difficult task due to the element of uncertainty. Hence, proper control and monitoring of the workforce needs to be ensured.

Role of spare parts:
A good maintenance management system requires appropriate tools. So, the system should have good quality tools available in the required quantities to ensure the proper function of the maintenance work.

Training of maintenance workforce:
Training of the workforce must be an integral part of any good maintenance management system. Training helps the workforce learn about modern techniques, and recent trends in maintenance and to chalk out a strategy to meet the growing demands of the industry.

1.8 Maintenance Planning

Planning maintenance jobs basically deals with answering two questions: 'What' and 'How' of the job, specifically, 'What activities are to be done?' and 'How those jobs and activities are to be done?' While answering these two questions, other supplementary questions are to be answered, e.g., 'Where the jobs are to be done?' and 'Why the job is to be done?', etc., as these will help in developing the 'What' and 'How' of the job. It is very essential that engineering knowledge must be extensively applied to maintenance jobs for the development of appropriate job plans using the most suitable techniques, tools, materials, special facilities, etc. (Braglia *et al.*, 2019). As job planning forms the basic foundations over which the efficiency and cost of actions depend, the persons responsible for job planning should have adequate capabilities, such as knowledge about jobs and available techniques, facilities and resources, analytical ability, conceptual logical ability, and judgmental courage.

Effective planning and scheduling significantly contribute to the following:

- Reduced maintenance cost.
- Improved utilization of the maintenance workforce by reducing delays and interruptions.

- Improved quality of maintenance work by adopting the best methods and procedures and assigning the most qualified workers for the job.

Planning Objectives

- Minimizing the idle time of maintenance workers.
- Maximizing the efficient use of work time, material, and equipment.
- Maintaining the operating equipment at a level responsive to the needs of production in terms of delivery schedule and quality.

Basic Levels of Planning Process (Depend on The Planning Horizon)

1. Long-range planning: It covers a period of three to five years and sets plans for future activities and long-range improvement.
2. Medium-range planning: It covers a period of one month to one year.
3. Short-rang planning: It covers a period of one day to one week. It focuses on the determination of all the elements required to perform maintenance tasks in advance.

1. Long Range Planning

It needs to utilize the following:

1. Sound forecasting techniques to estimate the maintenance load.
2. Reliable job standards times to estimate staffing requirements.
3. Aggregate planning tools such as linear programming to determine resource requirements. sets plans for future activities and long-range improvement.

2. Medium-Range Planning

- Specify how the maintenance workers will operate.
- Provide details of major overhauls, construction jobs, preventive maintenance plans, and plant shutdowns.
- Balance the need for staffing over the period covered.

- Estimate required spare parts and material acquisition.

3. Short-Range Planning

    - Focus on determining all the elements required to perform maintenance tasks in advance.
    - Assess required maintenance tools and skills needed for efficient maintenance of equipment.
    - Assess skills required for maintenance personnel.
    - Review personnel transfers to and from maintenance organizations assessing and reporting safety hazards associated with the maintenance of equipment.
    - Assess skills required for maintenance personnel.
    - Review personnel transfers to and from maintenance organizations.
    - Assess and report safety hazards associated with the maintenance of equipment.

1.9 Management Organization and Structures

- Manning Levels

Manning levels will be dependent upon factors such as asset numbers and types, skill levels, out-sourcing levels, productivity levels, or rosters.

- Technicians

The primary technician skill groups are electricians and mechanical fitters. There may be boilermakers and welders depending on the site and the need. Such staff are responsible for the physical implementation of the maintenance task upon an asset. Remotely-located sites, where the contractors and services have limited availability, will usually use more in-house staff.

- Non-Technical Assistants

Unskilled, non-technical assistants or laborers are also used to assist the skilled staff and/or perform non-technical tasks. Often, operations staff are used for this assistance.

- Procurement and Inventory Management

Often, the engineering department manages the store's inventory and procurement functions, although at some sites the procurement function is done by the finance department. Procurement management is responsible for maximizing value when purchasing items and ensuring that required items are on hand when needed. Store management is responsible for receiving and issuing items, stock control, store housekeeping, and also ensuring that required items are on hand when needed.

- Planning and Maintenance Data Management

Preventive maintenance practices require considerable planning and scheduling of related activities. There needs to be some capacity within the department to plan and monitor jobs and activities effectively.

- Supervision

Front-line supervision is required to manage the day-to-day activities during a shift, such as timekeeping, day-to-day job planning, liaising with the operations department on a shift basis, allocating jobs to staff, monitoring workflow, coaching and developing, and shift reporting. In the future, it is likely that more responsibility will be devolved to staff (tradespeople/technicians, etc.).

- Maintenance Management

Maintenance management is responsible for managing the department and takes overall responsibility for an asset's lifecycle during its operating life (Alnaggar and Pitt, 2019). Apart from overseeing the maintenance staff, maintenance management usually develops and manages the maintenance budgets, sets targets, makes decisions about what needs to be done, and looks at longer-term strategies.

- Engineering Support

There will usually be people within the engineering management group who can provide engineering support and technical assistance in mechanical and electrical disciplines. They will assist with day-to-day issues as well as project work.

- Quality Control

Often performed by supervision or could be done by peers. However, quality control of completed work is often overlooked.

- Training

All personnel should be trained and competent as appropriate. Management and supervision are responsible for assessing training needs. Often, adequate training is overlooked. Training should be given by in-house or external people depending on requirements.

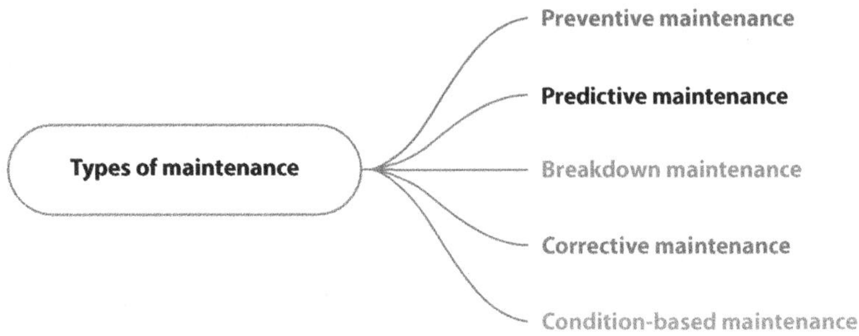

Figure 1.2 Types of maintenance.

1.10 Types of Maintenance (Figure 1.2)

The design life of most equipment requires periodic maintenance. Belts need adjustment, alignment needs to be maintained, proper lubrication on rotating equipment is required, and so on (Mustafa, 2021). In some cases, certain components need replacement, *e.g.*, a wheel bearing on a motor vehicle, to ensure the main piece of equipment (in this case a car) lasts for its design life. Different approaches have been developed to know how maintenance can be performed to ensure equipment reaches or exceeds its design life. In addition to waiting for a piece of equipment to fail (reactive maintenance), the other approaches are preventive maintenance, predictive maintenance, or reliability-centered maintenance (Poór *et al.*, 2019).

1.10.1 Breakdown (Reactive) Maintenance

Breakdown maintenance is basically the 'run it till it breaks' maintenance mode. No actions or efforts are taken to maintain the equipment as the designer originally intended to ensure design life is reached (Lucke, 2022). Recent studies indicate that this is still the predominant mode of maintenance. The advantages of breakdown maintenance can be viewed as a double-edged sword. If we are dealing with new equipment, we can expect minimal incidents of failure. If our maintenance program is purely reactive, we will not expend manpower or incur capital costs until something breaks. Since we do not see any associated maintenance cost, we could view this period as saving money. In reality, during the time we believe we are saving maintenance and capital costs, we are really spending more money than we would have under a different maintenance approach. We are spending more money associated with capital costs because, while waiting for the equipment to break, we are shortening the life of the equipment resulting in more frequent replacement. We may incur costs upon failure of a primary device associated with its failure causing the failure of a secondary device. This is an increased cost that we would not have experienced if our maintenance program had been more proactive. Our labor cost associated with the repair will probably be higher than normal because the failure will most likely require more extensive repairs than would have been required if the piece of equipment had not been run to failure. Chances are the piece of equipment will fail during off hours or close to the end of the normal workday. If it is a critical piece of equipment that needs to be back online quickly, we will have to pay maintenance overtime costs. Since we expect to run equipment to failure, we will require a large material inventory of repair parts. This is a cost we could minimize under a different maintenance strategy (Mishra and Tyagi, 2022).

Advantages

1. Involves low-cost investment for maintenance.
2. Less staff is required.

Disadvantages

1. Increased cost due to unplanned downtime of equipment.
2. Increased labor cost, especially if overtime is needed.
3. Cost involved with repair or replacement of equipment.
4. Possible secondary equipment or process damage from equipment failure.
5. Inefficient use of staff resources.

1.10.2 Preventive Maintenance

Preventive maintenance can be defined as, "Actions performed on a time or machine-run-based schedule that detect, preclude, or mitigate degradation of a component or system with the aim of sustaining or extending its useful life through controlling degradation to an acceptable level." Preventive maintenance is a means to increase the reliability of their equipment (Okpalaobi *et al.*, 2022). By simply expending the necessary resources to conduct maintenance activities intended by the equipment designer, equipment life is extended and its reliability is increased. In addition to an increase in reliability, a lot of amounts will be saved over that of a program just using reactive maintenance. Studies indicate that these savings can amount to as much as 12% to 18% on average.

Advantages

1. Cost-effective in many capital-intensive processes.
2. Flexibility allows for the adjustment of maintenance periodicity.
3. Increased component life cycle.
4. Energy savings.
5. Reduced equipment or process failure.
6. Estimated 12% to 18% cost savings over reactive maintenance program.

Disadvantages

1. Catastrophic failures are still likely to occur.
2. Labor intensive.
3. Includes performance of unneeded maintenance.
4. Potential for incidental damage to components in conducting unneeded maintenance.

Depending on the facility's current maintenance practices, present equipment reliability, and facility downtime, there is little doubt that many facilities purely reliant on reactive maintenance could save much more than 18% by instituting a proper preventive maintenance program. While preventive maintenance is not the optimal maintenance program, it does have several advantages over that of a purely reactive program. By performing the preventive maintenance as the equipment designer envisioned, we will extend the life of the equipment closer to design.

Preventive maintenance (lubrication, filter change, etc.) will generally run the equipment more efficiently resulting in dollar savings. While we will not prevent catastrophic equipment failures, we will decrease the number of failures. Minimizing failures translates into maintenance and capital cost savings (Karki *et al.*, 2022).

1.10.3 Predictive Maintenance

Predictive maintenance can be defined as "Measurements that detect the onset of a degradation mechanism, thereby allowing causal stressors to be eliminated or controlled prior to any significant deterioration in the component's physical state. Results indicate current and future functional capability" (Fermín-Cueto *et al.*, 2020).

Basically, predictive maintenance differs from preventive maintenance by basing maintenance needs on the actual condition of the machine rather than on some preset schedule. Preventive maintenance is time-based. Activities such as changing lubricant are based on time, like calendar time or equipment run time. For example, most people change the oil in their vehicles every 3,000 to 5,000 miles traveled. This is effectively basing the oil change needs on equipment run time. No concern is given to the actual condition and performance capability of the oil. It is changed because it is time. This methodology would be analogous to a preventive maintenance task. If, on the other hand, the operator of the car discounted the vehicle run time and had the oil analyzed at some periodicity to determine its actual condition and lubrication properties, he may be able to extend the oil change until the vehicle had traveled 10,000 miles. This is the fundamental difference between predictive maintenance and preventive maintenance, whereby predictive maintenance is used to define needed maintenance tasks based on quantified material/equipment condition.

There are many advantages to predictive maintenance. A well-orchestrated predictive maintenance program will eliminate catastrophic equipment failures (Dalzochio *et al.*, 2020). A schedule of maintenance activities can be made to minimize or delete overtime costs. It is possible to minimize inventory and order parts, as required, well ahead of time to support the downstream maintenance needs and optimize the operation of the equipment, saving energy costs and increasing plant reliability. Past studies have estimated that a properly functioning predictive maintenance program can provide savings of 8% to 12% over a program utilizing preventive maintenance alone. Depending on a facility's reliance on reactive maintenance and material condition, it could easily recognize savings opportunities exceeding 30% to 40%. Independent surveys indicate the following industrial average savings resultant from the initiation of a functional predictive maintenance program:

1. Return on investment—10 times
2. Reduction in maintenance costs—25% to 30%
3. Elimination of breakdowns—70% to 75%
4. Reduction in downtime—35% to 45%
5. Increase in production—20% to 25%.

Advantages

1. Increased component operational life/availability.
2. Allows for pre-emptive corrective actions.

3. Decrease in equipment or process downtime.
4. Decrease in costs for parts and labor.
5. Better product quality.
6. Improved worker and environmental safety.
7. Improved worker morale.
8. Energy savings.
9. Estimated 8% to 12% cost savings over preventive maintenance program.

Disadvantages

1. Increased investment in diagnostic equipment.
2. Increased investment in staff training.
3. Savings potential not readily seen by management.

1.10.4 Corrective Maintenance

Corrective maintenance is an important part of effective maintenance operations that can help improve the reliability, safety, and performance of your equipment. This guide helps you to understand the basics of corrective maintenance including how it differs from preventive maintenance what types of failure occur, and how you can create a corrective maintenance plan for your organization. Improve your maintenance operations with this guide on the fundamentals of corrective maintenance (Ramere and Laseinde, 2021).

Corrective maintenance is the process of troubleshooting and repairing equipment failures. It is typically applied after preventive maintenance has failed, during a breakdown, or in an emergency situation. Corrective maintenance seeks to identify and fix the root cause of an issue, preventing the same problem from occurring again in the future. Correctional maintenance refers to the actions done to rectify the defects noticed in equipment, whether they occur before or after a breakdown of the equipment.

Corrective maintenance is most effective when the root cause of an issue can be identified quickly and efficiently. Diagnostic tools such as vibration analysis, thermography, and oil analysis help technicians identify problems quickly before they can become catastrophic failures. It is important to work with skilled personnel to analyze the data captured in order to find a reliable solution that will stand the test of time.

Preventive maintenance is essential in ensuring long-term reliability, but there are several important benefits of corrective maintenance.

This type of maintenance allows technicians to fix specific issues without having to perform a full overhaul or replace the entire system. It also minimizes downtimes, reduces the overall cost of maintaining equipment, and helps ensure that machines or processes are operating in their optimal condition (Shamayleh *et al.*, 2020).

1.10.5 Condition-Based Maintenance

Discover the power of condition-based maintenance, and how it can benefit your business when it comes to reducing machine downtime. Get the facts now. Condition-based maintenance is a sophisticated strategy that helps businesses reduce machine downtime by responding to changing conditions in real time. Learn how this approach can help you cut costs, improve safety, and maximize equipment performance (Ganesh *et al.*, 2020).

Implementing condition-based maintenance helps businesses recognize problems before they lead to an equipment shutdown and maximize the cost efficiency of their operations. Companies can also reduce manpower costs by having less labor-intensive procedures in place. Additionally, it helps identify and prevent potentially hazardous mechanical failures as well as speed up turnaround times and increase productivity overall.

Concept of Reliability in Maintenance

Reliability is the probability of survival under a given operating environment. For example, the time between consecutive failures of a refrigerator where continuous working is required is a measure of its reliability. If this time is longer, the product is said to have high reliability (Woo and O'Neal, 2021).

In a textile mill, generally, the light is maintained at a minimum specified level. To achieve this, let us assume that there are 100 bulbs in use and the guaranteed lifetime of each individual bulbs is 5,000 hours. If we collect statistics about the number of bulbs that survived till 5,000 hours, we can compute the reliability of the bulbs. In this case,

Reliability = Failure rate = Number of bulbs survived till the specified time limit/Number of bulbs used.

If the number of bulbs survived till 5,000 hours is 80, then we can say that the reliability is 0.8 (i.e., 80/100)

The reliability of railway signaling systems, aircraft, and power plants are some of the interesting examples for demonstrating the reliability concept. In these cases, a failure will lead to a heavy penalty.

The concept of reliability can be matched with the systems concept. Generally, products/equipment will have many components which may function with serial or parallel relationships. So, the individual component's reliability affects the reliability of the product. Hence, enough attention must be given at the design stage such that the product's reliability is maximized. The cost of maintenance is also to be considered along with the reliability while improving it.

Reliability Improvement

The reliability of a system/product depends on many factors. So, we should concentrate at the grassroots level to improve the product's reliability.

Some of the ways of improving system reliability are listed below:

- Improved design of components
- Simplification of product structure
- Usage of better production equipment
- Better quality standards
- Better testing standards
- Sufficient number of stand-by units
- Usage of preventive maintenance at appropriate times when necessary

1.11 Economics of Maintenance

The concept of preventive maintenance is presently accepted in most industries to achieve their planned production. Both production and maintenance have the same objective–to produce a quality product with maximum efficiency and minimum cost. Theoretically, maintenance should aim at keeping the machines and other facilities in a condition that allows them to be used without any interruption and at their maximum profit-making capacity. Such a situation can be assured in practice only if machines are frequently replaced or if a stand-by machine is maintained which can be put into operation as and when the original unit is stopped for checks, repairs, and component replacements. But, no industrial unit can possibly afford to throw away its capital resources by replacing machines frequently, nor can a company block its money in equipment that will be only partially utilized by having stand-bys. An organization has, under such circumstances, to accept a certain loss in the productive capacity of its investments to enable maintenance, examine the various equipment, repair if not in order, and put them back in a condition as required by the users.

Analysis of Maintenance Cost
Downtime of equipment costs money to the company, which is made up of maintenance labor spent, materials and spares consumed, and supervision exercised. Loss of revenue due to downtime because of maintenance, however, provides savings to the organization through the extension of the useful life of plants and facilities and through maintaining them at the optimal standards of operation. These savings consist of a reduction in wastage and scrapping of materials, a higher level of employee morale and safety, minimization of process time, and postponement of the necessity of investment for capital replacement.

Management of maintenance has, therefore, to concern itself with the balancing of costs against gains so as to evolve the most suitable policies and determine the maintenance effort required, which as said earlier, is a function of cost and it can vary from industry to industry. In establishing a systematic maintenance system, the following stages are normally gone through.

0 stage - Only break-down maintenance
1st stage - Breakdown maintenance + a certain amount of cleaning and lubrication
2nd stage - Breakdown maintenance + planned lubrication and inspection
3rd stage - In addition to degree 2, preventive replacement of spares (renewal) is carried out
4th stage - In addition to degree 3, there are periodic maintenance schedules including overhauls
5th stage - In addition to degree 4, predictive maintenance techniques are adopted.

From the above, it can be assumed that as the degree of maintenance effort increases, the number of breakdowns decreases, while, on the other hand, maintenance costs also increase. Hence, there must be some compromise to achieve the optimum maintenance effort.

1.12 Maintenance Scheduling

Scheduling is the function of coordinating all of the logistical issues around the issues regarding the execution phase of the work. Scheduled maintenance jobs basically deal with answering two questions—'Who' and 'When' of the job, i.e., "Who would do the job?" and "When will the job be started and done?". Effective scheduling essentially needs realistic thinking based on substantial data and records. The majority of scheduling work needs to occur in areas such as overhead labor hours, safety and toolbox meetings, break times, training times, the addition of corrective and approved improvement actions as dictated by the prioritization system and operations plan, etc.

Requirements for Schedulers

A scheduler should also have knowledge about the job, techniques, facilities, analytical ability, and judgmental courage. The scheduler must obtain information about the following facts before starting their job:

1. Manpower availability by trade, location, shift, crew arrangement, permissible overtime limit, etc.
2. Man-hour backlog on current or unfinished jobs.
3. Availability of the equipment or area where the work has to be performed.
4. Availability of proper tools, tackles, spares, consumables, structural, and other required materials.
5. Availability of external manpower and their capabilities; these may be from other shops/departments of the plant or contractors (local, nearby, ancillary, etc.).
6. Availability of special equipment, jigs/fixtures, special lifting and handling facilities, cranes, etc. This should also include labor- and time-saving devices like pneumatic hammers, excavators, etc.
7. Starting date of the job; also often, the completion time of the total job is predetermined, and in that case, resources are to be arranged accordingly.
8. Past schedules and charts (updated) if the same job has been done earlier, etc.

1.13 Conclusion

We should look at maintenance as an integral part of production and apply the respective principles of production planning, work-study, stock control, etc., for the efficient fulfillment of this function.

Peter Drucker said, "In third world societies, the modern technology demand is a radical break with the social and cultural traditions and produces a fundamental crisis in the society." Specific to the Indian industry, the factors leading to such a crisis are:

- Economic dependency: In developing countries, mechanization is highly linked to foreign aid given by the donor countries, as well as the equipment supplied by the collaborators. Due to this and the limited foreign exchange situation, the Indian economy depends on the proper maintenance of the existing machinery and equipment.
- Technical dependency: This follows the same argument as above. The more sophisticated the machinery is, the more we have to depend on the supplying country for maintenance of the plant and equipment.
- Social change: Imported mechanized technology is basically the industrial culture and philosophy of Western countries. The fact that India has a rural-oriented society necessitates a different approach to education and training.

Maintenance plays a crucial role in asset management and operational performance. By adopting preventive, corrective, and predictive maintenance practices, organizations can enhance equipment reliability, extend asset lifespan, reduce costs, and improve overall productivity. Striking the right balance between maintenance efforts and resource allocation is key to achieving optimal maintenance outcomes.

References

Alnaggar, A. and Pitt, M., Towards a conceptual framework to manage BIM/COBie asset data using a standard project management methodology. *J. Facil Manage.*, 17, 2, 175–187, 2019, Emerald Publishing Limited.

Braglia, M., Castellano, D., Gallo, M., A novel operational approach to equipment maintenance: TPM and RCM jointly at work Vol. *J. Qual. Maint. Eng.*, 25, 4, 612–634, 2019, Emerald Publishing Limited.

Dalzochio, J., Kunst, R., Pignaton, E., Binotto, A., Sanyal, S., Favilla, J., Barbosa, J., Machine learning and reasoning for predictive maintenance in Industry 4.0: Current status and challenges. *Comput. Ind.*, 123, 103298, 2020.

Fermín-Cueto, P., McTurk, E., Allerhand, M., Medina-Lopez, E., Anjos, M.F., Sylvester, J., dos Reis, G., Identification and machine learning prediction of knee-point and knee-onset in capacity degradation curves of lithium-ion cells, in: *Energy and AI*, vol. 1, p. 100006, Elsevier, 2020.

Ganesh, S., Su, Q., Pepka, N., Rentz, B., Vann, L., Yazdanpanah, N., O'Connor, T. *et al.*, Design of condition-based maintenance framework for process operations management in pharmaceutical continuous manufacturing. *Int. J. Pharm.*, 587, 119621, 2020, Elsevier.

Helmold, M. and Samara, W., *Progress in Performance Management*, Springer, 2019.

Jian, Z.M., Qian, G.A., Paolino, D.S., Tridello, A., Berto, F., Hong, Y.S., Crack initiation behavior and fatigue performance up to very-high-cycle regime of AlSi10Mg fabricated by selective laser melting with two powder sizes. *Int. J. Fatigue*, 143, 106013, Elsevier, Amsterdam, Netherlands, 2021.

Karki, B.R., Basnet, S., Xiang, J., Montoya, J., Porras, J., Digital maintenance and the functional blocks for sustainable asset maintenance service–A case study. *Digital Business*, 2, 100025, Elsevier, 2022.

Lai, C.T.A., Jiang, W., Jackson, P.R., Internet of Things enabling condition-based maintenance in elevators service. *J. Qual. Maint. Eng.*, 25, 563–588, Emerald Publishing Limited, 2019.

Lucke, E., Operation, maintenance, and turnarounds, in: *Life Cycle of a Process Plant*, pp. 157–186, Elsevier, Amsterdam, Netherlands, 2022.

Mishra, S. and Tyagi, A.K., The role of machine learning techniques in internet of things-based cloud applications, in: *Artificial Intelligence-Based Internet of Things Systems*, pp. 105–135, Springer, Cham, Switzerland, 2022.

Mustafa, M.N., Classification of maintenance techniques and diagnosing failures methods. *J. Phys. Conf. Ser.*, 2060, 012014, 2021. IOP Publishing.

Okpalaobi, S., Mbah, S.I., Ezeanolue, E.T., Total Productivity Management And Performance Of Dangote Cements In North Central Nigeria, 2022.

Pieterse, B., Agyekum, K., Manu, P., Mohandes, S.R., Cheung, C., Yunusa-Kaltungo, A., Examining critical project management skills for successful delivery of major maintenance projects: Insights from the United Kingdom energy sector. *Eng. Constr. Archit. Manage.*, Emerald Publishing Limited, No. ahead-of-print, 4, 1556–1584, 2024.

Poór, P., Basl, J., Zenisek, D., Predictive Maintenance 4.0 as next evolution step in industrial maintenance development. *2019 International Research Conference on Smart Computing and Systems Engineering (SCSE)*, IEEE, pp. 245–253, 2019.

Ramere, M.D. and Laseinde, O.T., Optimization of condition-based maintenance strategy prediction for aging automotive industrial equipment using FMEA. *Procedia Comput. Sci.*, 180, 229–238, 2021, Elsevier.

Shamayleh, A., Awad, M., Abdulla, A.O., Criticality-based reliability-centered maintenance for healthcare. *J. Qual. Maint. Eng.*, 26, 2, 311–334, 2020, Emerald Publishing Limited.

Stone, M., Aravopoulou, E., Ekinci, Y., Evans, G., Hobbs, M., Labib, A., Laughlin, P. *et al.*, Artificial intelligence (AI) in strategic marketing decision-making: a research agenda, in: *The Bottom Line*, vol. 33, pp. 183–200, Emerald Publishing Limited, Bingley, United Kingdom, 2020.

Woo, S. and O'Neal, D.L., Reliability design of mechanical systems such as compressor subjected to repetitive stresses. *Metals*, 11, 1261, MDPI, 2021.

2

Condition-Based Maintenance

Rajdeep Singh* **and Chandan Deep Singh**

Punjabi University, Patiala, Punjab, India

Abstract

Condition-based maintenance (CBM) is a proactive maintenance technique that uses real-time data (collected through sensors) to identify when an asset's performance or condition reaches an unsatisfactory level. By observing the state of an asset, a practice known as condition monitoring, maintenance professionals can identify when an asset is about to fail or has failed. With CBM, maintenance work is performed only when needed in response to the asset's real condition, preventing unnecessary maintenance tasks. Though condition-based maintenance and predictive maintenance (PdM) have some overlap and are often used interchangeably, they are not technically the same. CBM focuses on real-time asset performance and conditions and alerts you the exact moment monitored parameters are out of bounds. For example, a sensor reading taken from an oil pump may show a major drop in pressure, indicating that a component has failed. Predictive maintenance utilizes real-time asset data, like CBM, in addition to predictive analysis to determine when an asset will fail in the future. To continue our previous example, sensor readings from an oil pump may be used to forecast when an issue will occur or is beginning to form. As you can see, while both CBM and PdM use condition monitoring, the key difference is timing–what is the condition of an asset right now (CBM) versus what might the condition of the asset be in the future (PdM).

Keywords: Condition-based maintenance (CBM), predictive maintenance, preventive maintenance, CMMS

Introduction

Condition-based maintenance (CBM) is a maintenance strategy that monitors the actual condition of an asset to decide what maintenance needs to

**Corresponding author*: rajdeepsingh.87@gmail.com

Chandan Deep Singh, Davinder Singh, Kanwal Jit Singh and Harleen Kaur (eds.) Condition-Based Maintenance and Residual Life Prediction, (23–46) © 2025 Scrivener Publishing LLC

be done. CBM dictates that maintenance should only be performed when certain indicators show signs of decreasing performance or upcoming failure. For example, you might install a sensor that measures the vibrations of a certain piece of rotating equipment. Over time, that moving piece will degrade and start falling out of the alignment which will result in an increased amount of vibration. The installed sensor can then warn you when the amount of vibration crosses a set limit so you know that part should be replaced shortly. The main goal of condition-based maintenance is to help you optimize your maintenance resources by performing maintenance work only when needed [1]. For condition-based maintenance to be successful, there are several other elements of your maintenance operation that need to be in place. This includes having a scheduled maintenance strategy that allows you to inspect and spot anomalies in equipment and trigger timely follow-up work orders as well.

Motor vehicles come with manufacturer-recommended intervals for oil replacements. These intervals are based on manufacturers' analysis, years of performance data, and experience. However, this interval is based on an average or best guess rather than the actual condition of the oil in any specific vehicle. The idea behind condition-based maintenance is to replace the oil only when a replacement is needed, and not on a predetermined schedule. In the example of industrial equipment, oil analysis can perform an additional function too. By looking at the type, size, and shape of the metal particulates that are suspended in the oil, the health of the equipment it is lubricating can also be determined [2]. The equipment maintenance supervisor or facility manager will determine a set of condition-based maintenance (CBM) indicators. In the event that these CBM indicators are met, the maintenance personnel will have to proceed with the maintenance of the equipment or machinery.

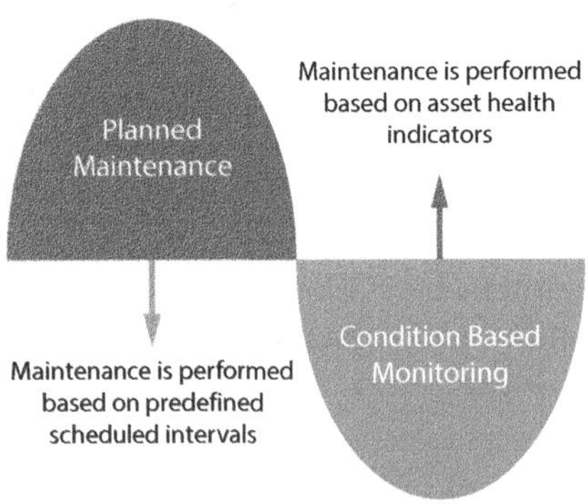

Figure 2.1 Planned maintenance vs. condition-based maintenance [2].

Condition-based Maintenance		
	Proactive Maintenance	**Predictive Maintenance**
What it looks for:	Root causes	Failure symptoms and faults
Example technologies employed:	Contaminant monitoring balance, alignment tools, viscosity and acid number monitoring	Wear debris analysis, vibration analysis, thermography, motor current analysis
Benefits sought:	Fault-free machine life extension	Early detection faults and failure
What's reduced?	Number of failures	Impact of failures (early detection)

Figure 2.2 Proactive maintenance vs. predictive maintenance [2].

The facility should have standard operating procedures (SOPs) to provide step-by-step instructions and guidelines on how the issue can be fixed by the maintenance personnel [3]. Unlike in planned scheduled maintenance (PM), where maintenance is performed based upon predefined scheduled intervals, condition-based maintenance is performed only after a decrease in the condition of the equipment has been observed. Compared with preventive maintenance, this increases the time between maintenance repairs, because maintenance is done on an as-needed basis.

Condition-Based Maintenance and Residual Life Prediction

Condition-based maintenance (CBM) is a maintenance strategy that monitors the actual condition of the asset to decide what maintenance needs to be done. CBM dictates that maintenance should only be performed when certain indicators show signs of decreasing performance or upcoming failure. Checking a machine for these indicators may include non-invasive measurements, visual inspection, performance data, and scheduled tests. Condition data can then be gathered at certain intervals, or continuously (as is done when a machine has internal sensors). Condition-based maintenance can be applied to both mission-critical and non-mission-critical assets.

Unlike in planned scheduled maintenance (PM), where maintenance is performed based upon predefined scheduled intervals, condition-based maintenance is performed only after a decrease in the condition of the equipment has been observed. Compared with preventive maintenance, this increases the time between maintenance repairs, because maintenance is done on an as-needed basis.

The goal of condition-based maintenance is to spot upcoming equipment failure so maintenance can be proactively scheduled when it is needed–and not before. Asset conditions need to trigger maintenance within a long enough time period before failure so work can be finished before the asset fails or performance falls below the optimal level.

Advantages

- CBM is performed while the asset is working, which lowers disruptions to normal operations.
- Reduces the cost of asset failures.
- Improves equipment reliability.
- Minimizes unscheduled downtime due to catastrophic failure.
- Minimizes time spent on maintenance.
- Minimizes overtime costs by scheduling the activities.
- Minimizes requirement for emergency spare parts.
- Optimizes maintenance intervals (more optimal than manufacturer recommendations).
- Improves worker safety.
- Reduces the chances of collateral damage to the system.

Disadvantages

- Condition monitoring test equipment is expensive to install, and databases cost money to analyze.
- Cost to train staff–you need a knowledgeable professional to analyze the data and perform the work.
- Fatigue or uniform wear failures are not easily detected with CBM measurements.
- Condition sensors may not survive in the operating environment.
- May require asset modifications to retrofit the system with sensors.
- Unpredictable maintenance periods.

Example of Condition-Based Maintenance

$$\text{CBM} = \text{Cost savings} + \text{Higher system reliability}$$

Condition-based maintenance allows preventive and corrective actions to be scheduled at the optimal time, thus reducing the total cost of ownership. Today, improvements in technology are making it easier to gather, store, and analyze data for CBM. In particular, CBM is highly effective where safety and reliability are the paramount concerns such as in the aircraft industry, semiconductor manufacturing, nuclear, oil and gas, etc.

Condition-based maintenance (CBM) is maintenance that arises when needed. This is performed when one of the indicators shows deterioration. This is applicable to critical systems that include fault reporting and active redundancy. It is a maintenance strategy that decides the sort of maintenance required. CBM commands that maintenance should be performed when some indicators show signs of upcoming failure or reducing performance. It is the process that predicts the present and future conditions of the machinery when in operation. It gathers information about the internal effects of the operating machine. Types of condition-based maintenance are (Randall, 2011):

(i) Vibration Analysis
Vibrations are always produced by machines, even when they are in good condition. This is due to periodic events in the machine's operation, such as rotating shafts, meshing gear teeth, rotating electric fields, and so on. The frequency of occurrence of such events often gives a direct indication of the source, and thus many powerful diagnostic techniques are based on frequency analysis.

(ii) Oil Analysis
This can be divided into three categories:
   (a) Spectrographic Oil Analysis Procedures (SOAP)
   (b) Chip Detectors
   (c) Ferrography

(iii) Performance Analysis
With different machines, performance analysis is a way of determining the correct functioning of the machine. As in gas turbine engines, there are many permanently mounted transducers for various process parameters like flow rates, temperatures, and pressures, and calculating various

efficiencies and comparing them with normal conditions is called flow path analysis. In modern internal combustion (IC) engine control systems, such as those for diesel locomotives.

(iv) Thermography
Thermography is used in quasi-static situations, such as with electrical switchboards, for the detection of local hot spots and faulty refractory linings in hot fluids containers.

(v) Infrared (IR) – IR cameras can be used to detect high temperature conditions in energized equipment.

(vi) Ultrasonic – Used for the detection of deep subsurface defects such as boat hull corrosion.

(vii) Acoustic – Used to detect gas, liquid, or vacuum leaks.

(viii) Electrical – Involves motor current readings using clamp on ammeters.

(ix) Operational performance – Sensors throughout a system to measure pressure, temperature, flow, etc.

Data can be collected from the system by two different methods:

- Spot readings can be performed at regular intervals using portable instruments.
- Sensors can be retrofitted to equipment or installed during manufacture for continuous data collection.

Example 1
One common application of condition-based maintenance is the use of pressure readings on piping systems. Pressure levels are constantly monitored, allowing the maintenance staff to identify when possible leaks may occur. Manufacturers who rely on pipe or liquid systems, such as food and beverage producers, use CBM to track pressure and diagnose problems.

Example 2
If you manage a fleet of vehicles, such as in the transport or construction industry, you may be used to scheduling an oil change based on a calendar schedule (such as every three months) or mileage (such as every 5,000 miles driven). But if you use CBM, your maintenance schedule may be informed by reading a gauge that monitors oil particles. This may mean changing the

oil every 8,000 miles instead of 5,000 miles or after four months instead of three months. Advanced monitoring would save costs on engine oil.

Example 3
Most buildings are upgrading to smart heating ventilation and air conditioning (HVAC) units that provide real-time temperature and humidity readings. This is a form of CBM that utilizes diagnostic algorithmic technology to adjust and control building temperature and alert management of any issues. This helps to prevent emergency shutdowns, or worse, fire damage.

Critical systems that require considerable upfront capital investment, or that could affect the quality of the product that is produced, need up-to-the-minute data collection. More expensive systems have built-in intelligence to self-monitor in real time. For example, sensors throughout an aircraft monitor numerous systems while inflight and on the ground to help identify issues before they become life-threatening. Typically, CBM is not used for non-critical systems where spot readings suffice.

Challenges of Condition-Based Maintenance

- Condition-based maintenance requires an investment in measuring equipment and staff upskilling, so the initial costs of implementation can be high.
- CBM introduces new techniques to do maintenance, which can be difficult to implement due to resistance within the organization.
- Older equipment can be difficult to retrofit with sensors and monitoring equipment, or can be difficult to access during production to spot measure.

With CBM in place, it still requires competence to turn performance information from a system into actionable proactive maintenance items.

CBM Components:
CBM components are an optimized mix of:

- Maintenance technologies (diagnostics, prognostics)
- Reliability-centered maintenance (RCM)-based processes
- Enablers (total asset visibility)

CBM Features:
Data acquisition may involve various types of information:

- Vibration
- Temperature
- Pressure
- Speed
- Voltage/current
- Stress/strain/shock
- Position
- Particulate count/composition

Feature extraction calculations may involve:

- Fast Fourier Transform
- Data filtering/smoothing
- Temperature/pressure ratio
- Efficiency
- Mass flow

- Detection algorithms alert users to potential problems and otherwise unknown failures.
- Diagnostic algorithms isolate failures to specific components or subsystems.
- Prognostic algorithms estimate remaining useful life based on past and future operational profiles and the physics of failure models.
- Supervisory reasoning algorithms reconcile conflicting information and provide recommendations such as:

- Inspections
- Repairs
- Parts ordering
- Equipment shutdown

Benefits:
The benefits of implementing CBM include:

- Increased system availability
- Increased system reliability

- Reduced maintenance costs
- Reduced inventories

How to Establish a Condition-Based Maintenance Program: Establishing a CBM program doesn't have to be difficult. Here's how to develop your CBM program in six simple steps:

- **Step 1: Select the equipment you want to monitor.** Give priority to the equipment that is key to business success, is high value, or has a longer lifespan.
- **Step 2: Identify the equipment's failure modes.** This includes all the known failure modes and probable ones. An RCM analysis can help with this. Always focus on the failure modes so that you can manage CBM strategies.
- **Step 3: Select the right condition-based monitoring system.** Ensure that the CBM system you choose can handle the failure modes identified in the previous step.
- **Step 4: Define your baseline limits for the system.** For the system to warn you when the equipment is starting to fail, you need to define the acceptable condition limits. As indicated in the P-F Curve, ensure that the limits set give you enough time (P-F interval) to take corrective actions.
- **Step 5: Establish your CBM program.** At this point, you should define tasks and assign responsibilities to your maintenance team. The focus should be on data collection and recording.
- **Step 6: Analyze the data and act as needed.** The data from your inspections and sensors should help you plot trends and schedule maintenance work whenever necessary.

Applications of Condition-Based Maintenance

CBM is a computerized maintenance management system (CMMS) software add-on that keeps tabs on the actual state of your critical assets by recording the output of any meters and gauges on that asset, making it easier to diagnose problems and prevent asset failure. Today's maintenance teams are using CBM to improve their efficiency and their organization's bottom line. The three most common applications of condition-based maintenance measure temperature, pressure, and oil viscosity.

A computerized maintenance management system, also known as Enterprise Asset Management (EAM) software, is designed to help schedule, plan, manage, and track maintenance activities associated with equipment, vehicles, or facilities. A CMMS solution provides a central storage location for the majority of maintenance data and information for assets, manages and controls your work and materials management/parts usage processes, and tracks maintenance activity over the lifecycle of an asset. All maintenance activities can be monitored and analyzed through robust CMMS reporting and dashboard tools.

Organizations across all industries can utilize a CMMS system. For manufacturers, a CMMS is used to manage maintenance for equipment and critical assets; for facilities managers, to manage maintenance for buildings and facilities; for service providers, to manage maintenance for clients, staff, tenants, and students; and for fleet operators, to management maintenance for vehicles and fleet equipment.

Computerized Maintenance Management Systems (CMMS)

Computerized maintenance management systems are well-adopted in industries in India. They may fetch many benefits which include increased productivity, reduced costs, and effective utilization of assets in manufacturing and service producers.

A CMMS is for any organization that is constantly working on reactive maintenance and unable to do preventive maintenance, frustrated with tracking and managing spare parts inventory, having difficulty with providing documentation for regulatory compliance, or wasting time on costly manual processes for tracking maintenance.

Organizations across all industries can utilize a CMMS system. For manufacturers, a CMMS is used to manage maintenance for equipment and critical assets; for facilities managers, to manage maintenance for buildings and facilities; for service providers, to manage maintenance for clients, staff, tenants, and students; and for fleet operators, to management maintenance for vehicles and fleet equipment.

Reliability-Centered Maintenance (RCM)

Reliability-centered maintenance, or RCM, is based on the concept of developing a customized setting in which all possible failure modes can be countered. The strategy focuses on finding the best solution to system failure by working on the type of failure in the individual machine. RCM is an overall best maintenance strategy that can be run throughout the product lifecycle to ensure product development. It can work very effectively in product development and establishment in the market (Wikstrom *et al.*,

2000). Any hurdles to product sales can be managed by using an effective maintenance plan.

The RCM approach is a powerful technique that can be used in developing the manufacturing performance in the industry. Even though the initial cost related to the implementation of the RCM model in the industry is high, the results are very effective in the long run. The approach can be helpful in developing a better system that yields greater productivity with fewer faults (Mkandawire, *et al.*, 2015). The overall quality of the system, its functions, and the product development are all increased with reliability enhancement.

Computer-Aided Facility Management (CAFM)

Computer-aided facilities management (CAFM) refers to the use of information technology to effectively manage physical facilities in various ways. This can include long-range facilities management reporting, as well as more direct systems that actively manage aspects of facilities, such as lighting or heating and air conditioning equipment.

Common features of computer-aided facilities management tools include systems that give information on floor plans and descriptions of the physical space, as well as reporting on energy consumption. These sorts of tools can also provide helpful equipment locations for a facility holding valuable or extensive business assets, such as machinery or hardware. Computer-aided facilities management tools can use extensive databases and visual modeling tools, along with geographic information systems or services, to give leaders and managers a bird's eye view of a particular facility or building.

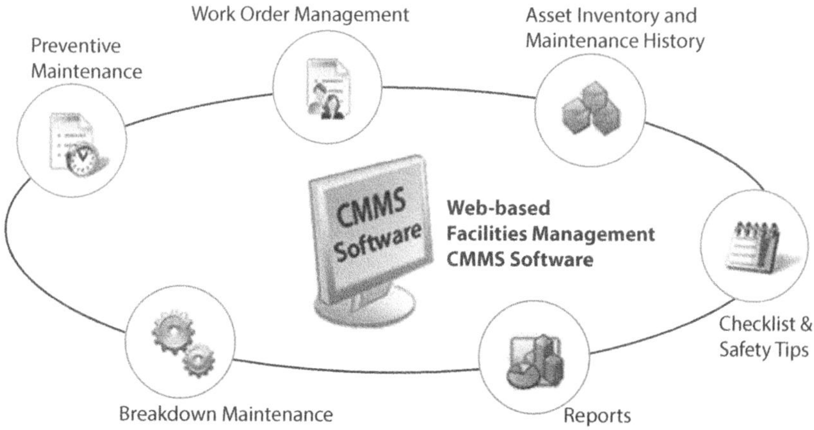

Figure 2.3 CMMS [2].

In addition to providing these sorts of physical tools, computer-aided facilities management systems can be part of long-term planning and auditing resources. For example, planners might use these sorts of tools to evaluate the depreciation of a facility or for other tax purposes. A wide range of CAFM solutions has become extremely valuable to those who are tasked with efficiently managing the physical locations of a business or other entity.

New Generation Strategic Asset Management (eXSAM)

Enterprise asset management (EAM) is the optimal lifecycle management of the physical assets of an organization. It covers subjects including the design, construction, commissioning, operations, maintenance, and decommissioning or replacement of plants, equipment, and facilities. "Enterprise" refers to the scope of the assets across departments, locations, facilities, and potentially, business units.

Enterprise asset management (EAM) is a broad term vendors use to describe software that provides managers with a way to view company-owned assets holistically. The goal is to enable managers to control and proactively optimize operations for quality and efficiency. In earlier years, EAM was simply called maintenance scheduling software. EAMs facilitate operations by automating requests for upgrades, regular maintenance, and decommissioning or replacement.

New Generation Strategic Asset Management (eXSAM) is a comprehensive strategic asset management solution that gives companies of all natures and sizes the power to manage, monitor, and control all of their assets and their unique attributes with one solution. eXsam streamlines and optimizes the management of all asset types, from complex mission-critical equipment, movable equipment, facilities, and plants, to people, livestock, entire mines, vehicles, IT assets, contracts, and fixed machinery. With eXsam, there's no need to deploy and manage different solutions to manage assets with unique attributes. Regardless of what a business's priorities are, or their asset mix, PricewaterhouseCoopers (PwC) Technology eXsam keeps it together.

1. Temperature Readings:

Temperature readings help monitor and control an asset's environment. Many assets can or do include sensors to gather temperature data, including infrared readings and thermal imaging. Also, organizations are using "smart" technology to help facility managers measure and control building temperature. Many universities, hotels, and hospitals are upgrading HVAC units to monitor temperature better. Windows and doors can also be outfitted with sensors. With these real-time readings, the HVAC can adjust for humidity and temperature, and alert staff to any issues. Machinery can also overheat, and computers must be continually cooled. Monitoring the temperature of these assets allows staff to prevent emergency shutdowns or even fire damage.

2. Pressure States:
Sensors can monitor pressure in equipment with piping or water systems. Monitoring pressure over time allows maintenance staff to predict when a pipe could fail and replace it beforehand. Instead of tracking down a leak after it happens, monitoring pressure indicates when and where a leak will likely happen before it does. Machinery in food and beverage production, as well as product manufacturing, relies on pipe or liquid systems. As competitive businesses are pushed to increase production while maintaining reliable operations, tracking pressure saves time in diagnosing problems. Maintenance staff can also keep parts at proper inventory levels for scheduled fixes.

3. Oil Viscosity:
A classic example of preventive maintenance (PM) is changing the oil in your car, either based on a calendar schedule (e.g., every three months) or based on mileage (e.g., every 5,000 miles). By adding condition-based maintenance to your preventive maintenance schedule, your maintenance team can monitor the oil particles to tell when to change the oil. Instead of 5,000 miles, the oil could last 8,000 miles. For construction or fleet vehicles, using CBM allows staff to schedule equipment maintenance and work more efficiently. In the long term, it also saves on oil and other parts [4].

The goal of condition-based maintenance is to uncover equipment failure before it happens, so maintenance can be done exactly when needed. Because CBM is based on collecting and analyzing data, it can be used to identify trends in asset performance and assess where an asset is in its lifecycle.

When Would You Use Condition-Based Maintenance?
For organizations looking for a more accurate way of scheduling maintenance based on real-time machine conditions, CBM allows you to monitor and identify machine degradation to enable proactive scheduling of maintenance activity when it is needed. CBM can result in a more effective maintenance strategy when compared with time- or calendar-based preventive maintenance techniques. Traditionally, it has been most often found in larger organizations with critical equipment to monitor and

significant downtime or safety issues to mitigate. However, advances in technology associated with Industry 4.0 have made CBM available to a much wider manufacturing community as an affordable and scalable solution for maintenance [12].

Condition-Based Maintenance (CBM) vs. Predictive Maintenance (PdM):

Condition-based maintenance (CBM) is a philosophy of guiding maintenance based on the condition of the equipment. PdM is a tool for testing the condition of equipment using trendable test results. Condition-based maintenance can be defined as equipment maintenance performed when certain indications imply performance degradation. This is often defined as the process being out of control or in chaos. It is different how the performance level is tested in condition-based maintenance as it is an umbrella term and is sometimes misused.

Predictive maintenance relies on advanced statistical methods, such as machine learning, to dynamically define when a machine is okay or need to be maintained. It looks at patterns across all sensors and makes one multivariate prediction model. The more data sources and data available the better the predictions. For this reason, predictive maintenance models do only get better at predicting future breakdowns over time. Predictive maintenance can find complex indications for breakdowns that would be nearly impossible for humans to spot. We have seen results predicting future breakdowns spanning from minutes to hours, depending on the quality of the data and the frequency at which the data is available (e.g., once an hour or once every 250 ms). But in some production settings where every breakdown can amount to millions in waste or produce security risks, even 5 minutes can be crucial [5].

Finally, predictive maintenance predicts future breakdowns by giving you a probability, whereas condition-based maintenance prevents additional breakdown costs by telling you something is wrong now. You can get started with both predictive maintenance and condition-based maintenance within a very short implementation period and start with zero data [6].

You might think it is easier to get started with a control chart than using machine learning; however, that is not necessarily the case. Both control charts and machine learning require time to prepare and clean the data, analyze the data, find the best methods, and so on. In condition-based maintenance, this involves testing which control chart, trimmed mean, dynamic models, etc., gives the best results, as well as testing in production if the method gives you too many alarms [7].

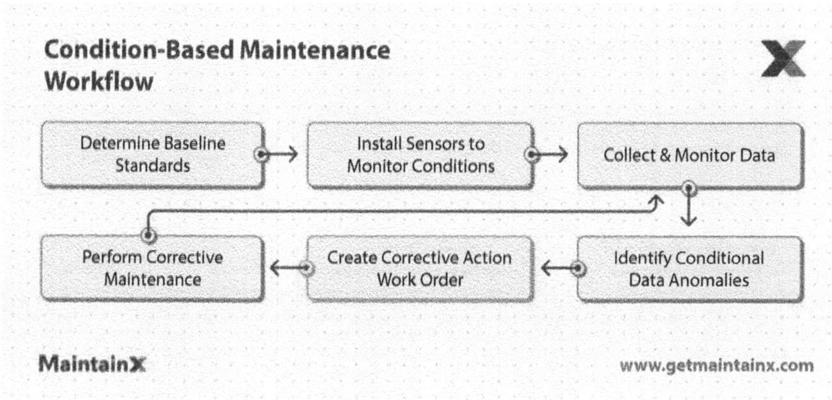

Figure 2.4 Condition-based maintenance workflow [12].

- **Condition-Based Maintenance:**
  1. A slightly higher increase in savings relative to costs.
  2. Condition-based maintenance requires some upfront investment in monitoring equipment and training. It also generally will have some recurring costs associated with software licenses and paying people to pay attention to the data and alarms (and some require humans to capture data manually).
  3. However, you will have a slightly higher proportional increase in savings as you ensure repair costs are only allocated to machines in need of it. This will provide greater peace of mind (rightfully so) that you won't have unexpected downtime of a machine providing greater savings.
- **Predictive Maintenance:**
  1. Higher increase in savings relative to cost increase.
  2. If you already have a condition-based maintenance strategy, then the only increase in costs will be for better and more data, and better analytics, which are relatively small compared to the initial investment in CBM.
  3. However, predictive maintenance will further increase machine uptime, as failures can be predicted and maintenance scheduled conveniently. Costs will also generally be less because the failure is caught even earlier [8].

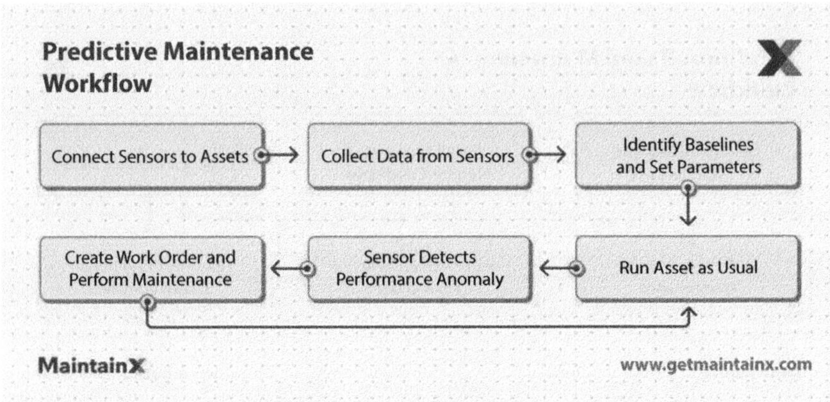

Figure 2.5 Predictive maintenance workflow [12].

Benefits and Pros of Condition-Based Maintenance:

- Increases asset life span
- Real-time (and often remote) indication of the health of an asset
- Maintenance only performed when needed
- Advanced warning provided to allow for convenient scheduling of maintenance

Drawbacks and Cons of Condition-Based Maintenance:

- Requires monitoring hardware and software to generate warnings which have upfront and recurring costs
- Some software may require the user to be specially trained to operate.

Benefits and Pros of Predictive Maintenance

- Increases asset lifespan.
- Provides a real-time indication of the health of an asset.
- Maintenance is only performed when needed.
- Provides advanced warning to allow for convenient scheduling of maintenance.

Drawbacks and Cons of Predictive Maintenance

- Requires monitoring hardware and software to generate warnings, which have upfront and recurring costs.
- Some software may require the user to be specially trained to operate [8].

Advantages and Disadvantages of Condition-Based Maintenance

Advantages:

- Minimizes the cost of asset failure.
- Improves asset productivity.
- Reduces downtime due to unplanned failure.
- Saves time spent on repair and maintenance after a breakdown.
- Enhanced efficiency in maintenance practices and management.
- Can be carried out while the asset is running, which reduces the number of disruptions to operations and keeps high production levels.
- Ensures an increase in the lifetime of any equipment.
- Reduces unscheduled downtime spent in the event of a major failure.
- Provides early detection of potential breaches, breaks, and wear on parts.
- Optimizes available resources, reducing unnecessary costs.
- Increases worker safety, preventing and avoiding work-related accidents.
- Prevents irreparable damage to industrial installations.

Various PdM Techniques

Most condition-monitoring devices and techniques include a sensor that must be installed on assets to collect data and send it to CMMS software. The software combines real-time sensor data with historical data and predictive models to provide maintenance teams with complete predictive analytics.

Sensor data is a central part of a predictive maintenance program internet of things (IoT) sensors can measure different aspects of pieces of equipment operations, including temperature, vibration, pressure, sounds, and more. Predictive maintenance is a technique that uses data analysis tools and techniques to detect anomalies in your operation and possible defects in equipment and processes so you can fix them before they result in failure. A report by McKinsey Global Institute estimates that current interest in linking physical assets to the digital world may actually still be understating its full potential. They project that this connectivity could generate up to $11 trillion annually in economic value by the year 2025 and that more than two-thirds of this value will be generated in a business-to-business setting.

The number of implementations and applications of predictive maintenance will increase as the cost of technology continues to drop, largely due to the switch from wire-based sensors to wireless ones. How does all this affect maintenance specifically? Well, predictive maintenance is changing the age-old way of doing things. It is no longer a case of "if it isn't broken don't touch it" but "keep it healthy so it doesn't break."

Optimal returns on the maintenance strategy require selecting the best combination of technologies and PdM techniques for the circumstance. Here are some of the main techniques:

1. Acoustic Analysis Monitoring:

This technique monitors machines' sound frequencies to detect problems in their technical performance and their sources. Most machines produce sound waves during operation, even if they are not detected by human ears. Acoustic sensors make it possible for technicians to hear friction and stress in rotating machinery, which indicates deterioration. Through acoustic technologies, maintenance teams can detect gas, liquid, or vacuum leaks in equipment on a sonic or ultrasonic level.

Even though this predictive maintenance method is considered less expensive than other technologies, such as ultrasonic technology, it has limited use. Acoustic analysis monitoring is especially useful for pipelines carrying liquids or gas.

Keep in mind that not all equipment necessarily needs a sensor to detect sound frequency changes. Some machines produce sounds loud enough for the human ear to detect changes. If that is the case, it might be a more cost-effective option. However, historical data would not be stored in the CMMS, and consequently, predictive models will not be created.

2. Infrared Thermography Analysis:

Infrared thermography leverages IR cameras to detect high temperatures in pieces of equipment. This imaging technology can be applied to a wide variety of equipment types to detect wear, rusting, delaminations, and disconnections that tend to go unnoticed by the naked eye.

Infrared cameras also allow maintenance teams to make surface temperature maps, which can be used, for instance, in the maintenance of heating systems. Sensors detect thermal anomalies in process systems relying on heat retention and transfer.

Worn equipment elements, including malfunctioning electrical circuits, typically emit heat that will display as a hotspot on a thermal image. This way, technicians can intervene with the right maintenance work and avoid equipment failure and even big accidents.

This nondestructive testing technology is considered one of the most versatile predictive maintenance technologies available. It can be used to analyze everything from individual parts of machinery to plant systems and even entire facilities.

3. PdM Oil Analysis:

Oil analysis is another noninvasive PdM technique with effective results. This method uses oil samples to determine equipment wear. It measures the particularities of oil samples, such as number and size, to determine machine deterioration.

Besides particle counts, some oil analysis tests can reveal the viscosity, the presence of water or wear metals, and the acid number or base number of the oil.

A hydraulic system, for instance, is composed of two main elements–the rotating parts and the lubrication part. As machines start declining in their operating conditions, the oil sample will contain byproducts of overheating and erosion. These particles reveal many potential issues and allow teams to come up with maintenance solutions before unplanned downtime happens.

With oil analysis, setting up the standards with historical data is easy for any organization operating in the oil and gas industry. The initial tests automatically set a baseline for the new machine and the results are stored in the database. If properly implemented, oil analysis yields a multitude of results that contribute to a successful predictive maintenance management plan.

4. Vibration Analysis:

This predictive maintenance technique uses built-in real-time sensors or handheld analyzers to monitor a machine's vibration. Since machines tend to have a regular vibration frequency during peak operation, it is possible to detect when there is an abnormal vibration pattern. Different vibration patterns occur when equipment parts, like bearings and shafts, start wearing off.

With continuous monitoring, CMMS software's algorithm compares the real-time data with failure modes and understands when maintenance tasks are needed. This predictive maintenance technique is employed primarily for high-speed rotating equipment. Among the issues that can be detected with vibration analysis are misalignment, bent shafts, unbalanced components, loose mechanical components, and motor problems.

Keep in mind that the cost associated with the implementation of this PdM approach is higher than the cost associated with other techniques.

Many heavy pieces of equipment have several joints. To keep the equipment safe and to keep employees safe, it is essential that vibration checking is done at regular intervals. This helps the organization determine whether asset joints are in good condition or not. For this purpose, real-time sensors are used. These sensors are helpful in detecting abnormal vibrations. Therefore, it is also an important predictive maintenance technique.

5. Motor Circuit Analysis:

Motor circuit analysis leverages electronic signature analysis (ESA) to identify issues and potential equipment failures based on electric motor components. Electronic signature analysis measures a motor's supply voltage and operating current to determine issues.

This technique helps maintenance teams get a complete picture of the electrical health of a critical equipment's motor system. It can recognize problems in the stator winding, issues in the bearing, rotor problems, problems in the coupling, abnormalities in the attached load, efficiency, system load, and others.

An advantage of this method is that motor circuit analyzer maintenance tools allow tests to be conducted in about two minutes and while the equipment are still running.

6. Streamline PdM Techniques With a CMMS:

Regardless of the PdM techniques, you will need to leverage your maintenance activities; investing in modern CMMS software will significantly increase productivity, cost-savings, automation, and return on investment (ROI). This is because it integrates all sensor data and organizes it into valuable predictive analytics.

Predictive maintenance technology is in constant evolution, so being on top of the best tools is crucial to building a strong and cost-effective maintenance strategy. Limble's Modular IoT approach allows you to easily implement a predictive maintenance strategy that fits your facility's needs [9]. If you want to get more information about how your specific organization can leverage predictive maintenance techniques, request a demo with one of our team members.

7. Real-Time Information:

When an organization uses real-time information, it is helpful in understanding performance while it is working. This allows the maintenance team to check if an asset is working well or needs some calibration adjustments.

If adjustments need to be made, then a checklist is created with a predictive maintenance solution, and this adjustment is made when asset maintenance is performed.

8. Key Performance Indicators List:

In key performance indicators, organizations consistently monitor assets so that the performance of a particular asset can be calculated. This helps get valuable data from the asset so that the organization can establish a bar for what to expect from an asset. Significant key performance indicators are unplanned downtime, asset failure or reactive maintenance done, maintenance expenses, and so on.

9. PdM in the Manufacturing Industry:

The benefits of the increased reliability that PdM offers are quite enticing and manufacturers are taking note. The same report mentioned above indicates that manufacturers' savings from using predictive maintenance could reach between $240 billion and $630 billion globally by 2025.

The everyday applications of predictive maintenance in manufacturing are numerous. Specific examples include:

- Using vibration sensors to identify patterns of fragile spindles in milling machines.
- Measuring the temperature differential upstream and downstream of the heat exchanger to identify the first signs of clogging.
- Monitoring robot CPU and housing temperatures, as well as positioning and overload errors, and using this data to estimate the health of a robot [10].

10. PdM in the Transportation Industry:

Connected transportation devices are on the rise as brands produce vehicles equipped with sensors that gather data and relay information about the machine's performance. Fleet managers can use this information to proactively plan for vehicle maintenance.

These practices have resulted in a measurable decrease in on-road breakdowns and spending, and have reduced technician diagnostic time. Fewer unplanned downtime, lower fuel and maintenance costs, and shorter servicing times are some of the reasons why fleets are moving towards predictive maintenance.

11. PdM in the Aviation Industry:
Interestingly, the origin of predictive maintenance can be traced to aircraft maintenance. Back in 1943, a British scientist named C.H. Waddington commented that "inspections tend to increase breakdowns" after he observed the maintenance activities of the Royal Air Force Coastal Command 502 Squadron. Instead, he recommended a process that was more "in-tune with the actual condition of the equipment."

Since then, PdM has evolved and is still used by aviation companies to reduce the costly effects of maintenance-related flight delays and cancellations. Predictive maintenance is also used in the industry to help airlines monitor engine health and performance during flights by measuring various temperature and vibration levels.

References

1. https://www.reliableplant.com/Authors/Detail/2367.
2. https://www.reliableplant.com/condition-based maintenance-31823.
3. https://www.fiixsoftware.com/maintenance-strategies/condition-based.
4. https://safetyculture.com/topics/condition-based-maintenance/.
5. https://mpulsesoftware.com/blog/maintenance-management/three-common-applications-condition-based-maintenance/.
6. eurospace.io/blog/2019/08/condition-based-maintenance-vs-predictive-maintenance/.
7. Rasay, F. and Zaremehrejerdi, Application of Multivariate Control Charts for Condition Based Maintenance. *Int. J. Eng.*, 31, 4, 597–604, 2017.
8. Liu, J. and Zhang, *An integrated model of statistical process control and condition-based maitnenance for deteriorating systems*, 2017.
9. https://blog.endaq.com/differences-between-condition-based-predictive-and-prescriptive-maintenance.
10. https://limblecmms.com/predictive-maintenance/predictive-maintenance-techniques/.
11. https://limblecmms.com/predictive-maintenance/#application-of-PdM-across-different-industries.

12. https://www.getmaintainx.com/learning-center/condition-based-maintenance/.
13. https://www.senseye.io/condition-based-maintenance.

3

Condition Monitoring

Harpreet Sharma[1*], Chandan Deep Singh[2] and Kanwal Jit Singh[3]

[1]*Yadavindra Department of Engineering, Guru Kashi Campus, Talwandi Sabo, Punjab, India*
[2]*Department of Mechanical Engineering, Punjabi University, Patiala, India*
[3]*School of Agriculture Sciences and Engineering, Maharaja Ranjit Singh Punjab Technical University, Bathinda, India*

Abstract
Condition monitoring is a crucial aspect of maintenance and reliability engineering that involves the continuous monitoring of equipment and systems to identify potential faults or abnormalities. By employing various techniques such as vibration analysis, thermal imaging, and oil analysis, condition monitoring enables proactive maintenance practices and helps in minimizing unexpected equipment failures, optimizing maintenance schedules, and reducing downtime. This chapter provides an overview of the importance of condition monitoring and its applications across various industries. It also highlights the benefits of implementing condition monitoring strategies and emphasizes the need for advanced monitoring technologies in the era of Industry 4.0.

***Keywords*:** Condition monitoring (CM), predictive maintenance (PdM), ultrasound testing, oil analysis, acoustic emission, infrared thermography, real-time monitoring

3.1 Introduction and Meaning

Condition monitoring is defined as measuring specific equipment parameters, such as vibrations in a machine, its temperature, or the condition of its oil, taking note of any significant changes that could be indicative of an impending failure. Continuously monitoring the condition of equipment

**Corresponding author*: harpreetsharma3211@gmail.com

Chandan Deep Singh, Davinder Singh, Kanwal Jit Singh and Harleen Kaur (eds.) Condition-Based Maintenance and Residual Life Prediction, (47–64) © 2025 Scrivener Publishing LLC

and taking note of any irregularities that would normally shorten an asset's lifespan allows maintenance or other preventive actions to be scheduled to address the issue(s) before they develop into more serious failures (Muthanandan and Nor, 2019).

Condition monitoring is a big component of predictive maintenance. The data collected from condition monitoring over time provides valuable information about the current and historical state of an asset. This evolution of a machine can be used to anticipate how the asset will perform over time and how it might degrade, allowing for the scheduling of maintenance based on these predictions. This is known as predictive maintenance–maintenance based on what failures may occur and what maintenance should be scheduled to prevent such failures from occurring.

Condition monitoring techniques are typically used on rotating equipment (gearboxes, reciprocating machines, centrifugal machines, etc.), backup or secondary systems, and other machinery such as compressors, pumps, electric motors, presses, and internal combustion engines.

There are two common methods used for condition monitoring:

- **Trend Monitoring:** Trend monitoring is the continuous, regular measurement and interpretation of data. It involves choosing a suitable and measurable indicator of machine or component deterioration and studying this trend to figure out when deterioration goes over a critical limit. For example, trend monitoring is used for routinely tracking airplane engine data to detect and diagnose abnormalities in engine performance, hopefully preventing secondary, more costly damage.

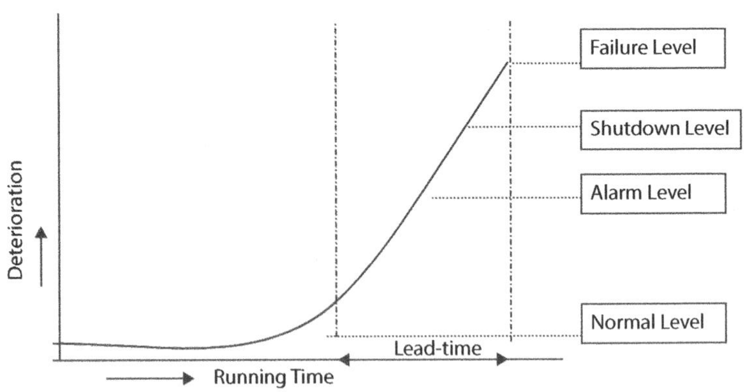

Figure 3.1 Trend monitoring.

The various states that the equipment is expected to undergo before breaking down are shown in Figure 3.1 above.

Normal Level
This is the state of the equipment in which it is running satisfactorily and production is within the normal range.

Alarm Level
This is the state that develops in the equipment during the early stages of a failure/fault and is noticed by some predetermined. For example, we may most likely experience sneezing, coughing, headaches, etc., before getting a fever. Low production, lower quality of the output, heavy noise, vibrations, bad smell, discoloration, etc., are some of the symptoms that indicate the alarm level of a failure or fault of the equipment.

Shutdown Level
If neglected even after reaching the alarm level, the fault grows and reaches a state where the equipment will not cooperate, or a considerable loss of output is observed. This state is called the shutdown level. This state is on the threshold of the breakdown level. The equipment should necessarily be closed or shut down and should not be run any longer in the interest of preventing further failure.

Breakdown Level
This is the ultimate and conclusive failure condition of the equipment, when outputceases or a catastrophe, even with potential damage, may occur.

Lead Time
This is the duration between the start of a failure and the breakdown in the process of deterioration of the equipment. In other words, it is the time gap between the last point of the normal level and the breakdown level of the equipment. Lead time is one of the main merits of using trend monitoring rather than the simple alarms or automatic shutdown devices used in permanent monitoring.

In most cases, both methods, condition checking and trend monitoring, are used together. With the use of sensors, automation of monitoring is possible. Permanent monitoring by such sensors can prevent catastrophic failures by alarming or stopping the machine automatically when the parameter (selected to indicate the machine condition) exceeds a predetermined level. Of course, the preselected warning/alarm levels, based on the experience of the users show a way to optimize the life of wearing machine parts. Thus, in condition monitoring, repairs are carried out only when the machine condition deteriorates to a predetermined level. In other words, repairs or replacement of parts take place only when it has definitely been proved that a fault exists. If left as 'unrepaired', this fault would result in unsatisfactory operation such as decreased production or catastrophic breakdown, with possible damage to other machine parts and disruption of production.

- **Condition Checking:** Condition checking involves taking a periodic check measurement with an appropriate indicator while a machine is running. The information from this method is then used to measure the condition of the machine at a given time. An example of condition checking could be using an oil sight glass, like a condition monitoring pod (CMP), to check the condition of a machine's lubricant in real time.

Condition monitoring via these two methods provides an inside look at how your machines and/or components are currently operating, and over time, offers a historical account of machine health.

Figure 3.2 shows a schematic layout of a condition monitoring system. In principle, there is no difference in the layout of the system regarding monitoring a process, an object, or the environment. First, it is necessary to clarify and determine the variables that should be measured. Afterwards, a suitable system should be chosen and installed. After recording, the values

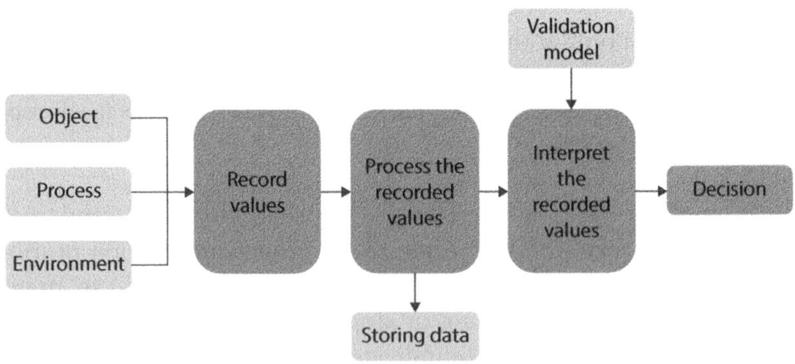

Figure 3.2 Schematic layout of a condition monitoring system.

should be processed and saved. Furthermore, after processing the values, they must be interpreted by an expert or technician with specialized knowledge. This requires having a suitable validation model. Based on the validation model or the specialized knowledge of the experts or technicians, a decision should be made.

3.2 Advantages of Condition Monitoring

As touched upon above, condition monitoring offers benefits for maintenance scheduling, operating costs, reduced downtime, and safety. It allows users to determine whether an asset is likely to break down, what will cause the problem, and when the failure may occur. This gives owners time to organize maintenance before the failure happens, avoiding unplanned downtime and allowing repairs to be set for a suitable time in a production schedule (Sharp *et al.*, 2022).

As such, the advantages of condition monitoring include:

1. *Avoid Unplanned Downtime*
Avoiding unplanned downtime offers a range of beneficial outcomes for an industrial environment. By removing unplanned downtimes, you prevent unplanned production delays that can lead to a loss of reputation if orders are not completed on time and the need to pay overtime to complete a job. In addition, there is no need to pay for an emergency call out to maintenance staff, which is more expensive than pre-scheduled maintenance.

2. *Protect Your Other Assets*
A machine failure can cause damage to other systems, compounding the problem further while also increasing repair or replacement costs. As a result, there will be no need to buy and store large numbers of replacement assets or parts in case of an unexpected failure. Instead, you will be forewarned of the need to replace a part and can order it as required in time for scheduled maintenance.

3. *Eliminate Unnecessary Maintenance to Maximize ROI*
Using condition monitoring as part of a predictive maintenance program can increase the return on investment (ROI) of mechanical assets. Preventive maintenance programs are set at regular intervals (i.e., annually) or after a set number of operating hours, irrespective of whether the maintenance is required or not. Through condition monitoring, it is possible to eliminate unnecessary maintenance and downtimes by only scheduling

repairs when required. This means that you can get more working value from each machine, reducing the total cost of ownership and maximizing the ROI for your equipment.

4. *More Efficient Maintenance*
Condition monitoring allows for more efficient maintenance by indicating where a fault may lie. This means that maintenance engineers do not need to check working components while locating the fault. This not only saves time and gets your maintenance completed faster, but also saves the cost of paying a maintenance engineer for wasted time.

5. *Improved Safety*
Checking and repairing machines before they break safeguards employee safety and helps ensure safer work practices. Condition monitoring means that owners can plan maintenance before failure occurs, which could pose a threat to employees working nearby.

6. *Improved Asset Efficiencies*
Condition monitoring can also improve the efficiency of assets. By having a record of which parts are running poorly, you can focus efficiency improvement efforts on those specific parts, thereby improving the overall capabilities of your equipment.

Downsides of Condition Monitoring
Despite the many advantages to be found with condition monitoring, there are also some drawbacks, particularly related to the initial setup of condition monitoring systems.

These drawbacks include:

1. *Installation Costs*
Condition monitoring equipment can be expensive to buy and install. In some instances, the monitoring systems may even require assets to be modified for retrofitting with sensors. There may also be additional sensor costs to account for operating environments, further increasing the price of installation.

Some asset owners may decide that certain items of equipment are not worth the investment and prefer to either schedule regular maintenance for easily repairable items or even run non-safety-critical items to failure.

2. *Operational Costs*
The ongoing data analysis required for condition monitoring will require either staff training to implement or the hiring of engineers with the

required knowledge and experience. These will both incur additional costs, although these can often be offset by the many financial benefits of condition monitoring (as shown above).

3. *Unpredictable Maintenance Scheduling*
While condition monitoring offers many advantages for maintenance scheduling, some operators prefer to simply schedule maintenance on a regular basis. For example, it may be preferable to service a machine every six months, rather than waiting for it to show signs of wear at an unexpected time. However, most modern condition monitoring techniques offer advanced warning of a potential failure, offering plenty of time to schedule maintenance (Liu and Zhang, 2020).

3.3 Condition Monitoring Applications

Condition monitoring applies to thousands of applications, but the best known include:

- Industrial Plant and Facilities of All Types: Gearboxes, uninterruptible power supply (UPS), alternating current (AC), electric motors, fans, pumps.
- Pulp and Paper: Blowers, conveyor belts, chippers, chip classifiers, refiners, pressure screens, screw conveyors, agitators, nip monitoring, felt rollers, etc.
- Iron and Steel: Raw material handling machines, conveyor belts, ship unloaders, galvanization plants, stack reclaimers, continuous casters, cranes, rolling mills, annealing machinery, and also pumps, fans, gearboxes, etc.
- Automotive: Wind tunnels, air handling units and pumps in paint shops, as well as presses and transfer presses, etc.
- Cement: Crushers, gearboxes, conveyor belts, separators, fans, raw mills, ball mills, elevators, and blowers.
- Power Generation Plants: gas turbines, steam turbines, water pumps, etc.

3.4 Four Pillars of Condition Monitoring

Condition monitoring, also known as Dynamic Predictive Maintenance, is composed of several functional aspects on which the maintenance is carried out. However there are four important schematic and functional steps observed in any implementation of condition monitoring activity. These functional aspects are often called the four pillars of condition monitoring. These are:

- Detection
- Diagnosis
- Prognosis
- Program

Detection

This pillar of Dynamic Predictive Maintenance (DPM) determines when a developing fault has just arisen. It focuses on answering the following 'When' questions:

(a) When might the fault have started?
(b) When (or how early) is the fault noticed?
(c) When will this fault result in a disaster if left unattended?

Diagnosis

Diagnosis concentrates on knowing the 'what' of the origin of the fault so that spare parts can be ordered.

(a) What is the origin of the fault?
(b) At what stage is the fault now?
(c) What is the growth rate of the fault?
(d) What are the components or parts that are suffering from this fault?

Prognosis

This pinpoints what could happen if the fault is left unattended, i.e., the FORECAST of after-effects. It is to estimate the severity of the fault, the possibility and/or frequency of its recurrence, operational problems, etc. Subsequent measurements will then establish the trend to indicate what may happen. This function puts the effort into answering the following queries:

(a) What is the severity of the fault?
(b) What could be the results of the fault?
(c) What could be the production loss?

(d) What is the cost of the fault in terms of hours of production and maintenance or repair?
(e) What are the side effects?
(f) What is the trend of the failure? (progressive/constant/ retrogressive)
(g) What is the lead time?

Program
This function is concerned with the preparation of a SCHEDULE to enable planned repairs. This function attempts to answer the following types of questions:

(a) When should the equipment be taken for maintenance or repair?
(b) What maintenance policy should be adopted?
(c) When should the restoring action start, and what is the likely time for its completion?
(d) Who will maintain or repair so as to restore the condition?
(e) What is the action plan?

3.5 Setting Up a Condition Monitoring (CM) Activity

There are a number of aspects to be taken into account if an industry is considering whether to apply CM to its plant, machinery, or equipment. The following suggested procedure may be helpful:

(a) The management should become familiar with the principles of CM and its benefits.
(b) Then they should consider various factors affecting the application of CM in relevance to their situation and also the actual merits/benefits that they intend to materialize.
(c) Then, the possible financial savings are to be assessed based on the general financial statistics of the organization.
(d) While making a general financial assessment, they should also identify what general or organizational arrangements appear to be appropriate.
(e) After considering various aspects (a) to (d), the management should discuss with their senior engineers concerned with the operation and maintenance of their plant and equipment. The engineers should be asked to make a more

precise estimate of the likely savings and to recommend an appropriate level of monitoring activity.

(f) If these more precise figures appear to be satisfactory, a decision should be made to proceed.

(g) A senior engineer should then be asked to survey the plant and equipment to recommend which items should be monitored and by what methods and then prepare cost estimates.

(h) A final decision should then be made on the actual expenditure and approval is to be given to order the necessary equipment or services and to train the required staff. CM can then be brought in a complete package.

(i) Management support and administrative work for setting up befitting procedures will be necessary during the first year of operation to ensure the smooth integration of CM into existing operations and the most effective utilization of machine data collected.

3.6 Condition Monitoring Types

There is a variety of machine condition monitoring types, each serving a different role in collecting data. Below are the most common types.

Online Condition Monitoring

Online Condition Monitoring is a type of Condition Monitoring that emphasizes the continuous monitoring of a machine or any production process. Online condition monitoring is a non-interrupted process. There is no word like shutdown in online condition monitoring. This helps in cost reduction and prevents unexpected costs. We cannot afford a loss in production and equipment failure in today's competitive age due to increasing demand. Read more about the difference between online and offline condition monitoring, see Shafiq *et al.* (2019). Figure 3.3 shows the evidence of Online condition monitoring.

Benefits of Online Condition Monitoring

There are several benefits to opting for online condition monitoring of machinery and technology in a plant/factory/establishment. Some are listed below:

- Online condition monitoring gives you timely and secure information about the machine's condition.

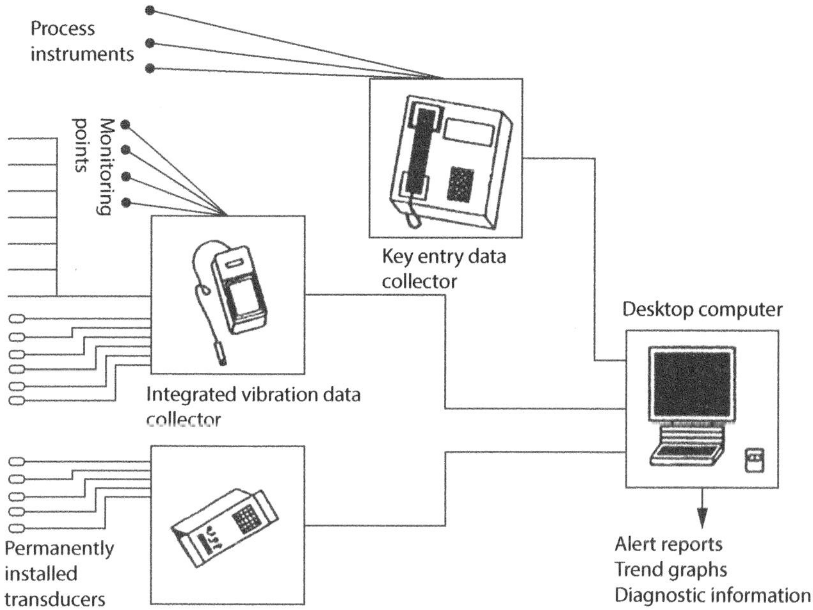

Figure 3.3 Online condition monitoring.

- It helps in pre-estimating the upcoming faults and figures out major ones.
- Online condition monitoring helps in reducing operating costs and total costs.
- You can measure the actual depreciation values of the assets.
- Online condition monitoring helps in reducing unplanned downtime.
- Reduction in maintenance cost.
- It helps in supporting ongoing reliability.
- It reduces risk and fewer inspections.
- Provides the surety of confined and remote assets.
- It helps you in drafting out hourly, daily, weekly, monthly, and quarterly statements even.
- It helps in providing direct access to the database.
- It makes you more efficient and compliant in all respects.
- Helpful during audits and in maintaining records.
- Scheduled or preventive maintenance reduces expenses and results in better equipment availability.
- It is an advanced and latest technique to monitor the conditions of machinery.
- Reduces the efforts for inspection and maintenance.

Offline Condition Monitoring
Offline condition monitoring is generally used for less critical or semi-critical assets where periodic scanning is good enough to observe the current condition. Offline monitoring is most commonly utilized with vibration analysis when a periodic check of the pulse will suffice for a less critical machine, and with oil analysis when oil is sent to a laboratory for testing. With offline oil analysis, some organizations use sampling kits to test certain oils' viscosity and water levels onsite for basic results. Semi-automated, in-house testing equipment is also available to test for things like wear metals, nitration, oxidation, and additive depletion (Chavan *et al.*, 2022).

Route-Based Condition Monitoring:
Route-based condition monitoring is a technique where a technician intermittently records data using a handheld device like an infrared imaging camera. Often, this method is employed to create a trend pattern and determine if more advanced analysis is needed (Tang, 2020).

3.7 Condition Monitoring Techniques

Condition monitoring techniques are standardized through International Organization for Standardization (ISO) and the American Society for Testing and Materials (ASTM). ASTM outlines a variety of standards, mostly dealing with condition monitoring for in-service lubricants, while ISO standards 13372, 18436, 17359, and 13381 (among others) specify the guidelines for condition monitoring and diagnostics of machines (https://www.sciencedirect.com/science/article/pii/S2214785321057448).

Figure 3.4 Vibration analysis.

Below are the most common techniques used for gathering data on the current condition of machinery.

- **Vibration Analysis:** Vibration analysis (Figure 3.4) is a process for measuring the vibration levels and frequencies of machinery and then using that information to analyze the machine's health. Vibration analysis can detect issues like imbalance, bearing failures, mechanical looseness, misalignment, resonance and natural frequencies, electrical motor faults, bent shafts, and even cavitation. It's been estimated that vibration warnings can provide up to three months of lead time before an actual failure occurs. The link above provides an in-depth look into vibration analysis (Nithin *et al.*, 2022).
- **Oil Analysis:** Oil analysis (Figure 3.5) is used to routinely analyze the health of machinery lubricants, oil contamination, and

Figure 3.5 Oil analysis.

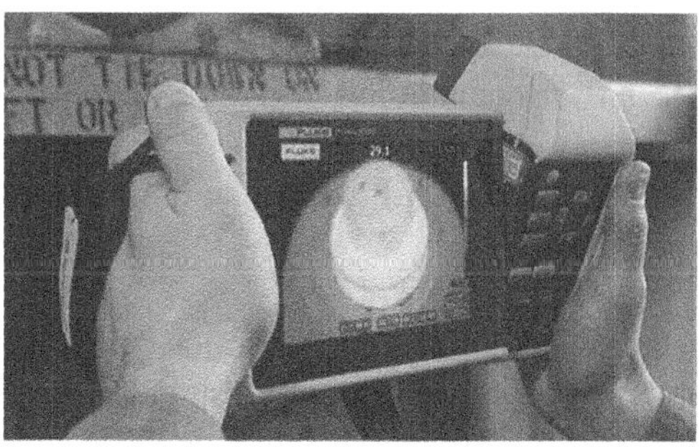

Figure 3.6 Infrared thermography.

machine wear. The oil analysis process includes moisture analysis, particle counting, elemental analysis, acid/base numbers, measuring viscosity, and using Fourier transform infrared (FTIR) spectroscopy to determine several other parameters. For example, a spectrographic oil analysis test can break down the chemical composition of oil to predict possible issues.

High levels of silicon and aluminum in oil reveal it is contaminated with dirt or grit (aluminum silicates), while high levels of iron indicate wear components. The link in this section offers an overview of oil analysis, including what it measures and what to look for in an oil analysis report (Soliman, 2020).

- **Infrared Thermography:** Infrared thermography (Figure 3.6) is the process of using a thermal imager to detect radiation (heat) emitting from an object, converting it to temperature, and then displaying that temperature distribution in an image. It is commonly used to monitor electrical and mechanical conditions of motors, bearings (abnormal friction), refractory insulation, and locating gas, liquids, and sludge levels. The primary goal of infrared thermography is to ensure machinery is running normally by detecting abnormal heat patterns within a machine that could indicate a defect or inefficiency. The link above explains the types of infrared thermometers and cameras, how to use and assess infrared thermography, and more (ElMasry *et al.*, 2020).
- **Ultrasound:** Ultrasound testing is useful for low- and high-speed mechanical applications, and high-pressure fluid situations. For example, a digital ultrasonic meter measures high-frequency signals emitted from bearings and displays those signals in real-time in decibels per microvolt (dBuV). Over time, this metric creates a historical picture to predict increases in friction, rubbing, and other bearing defects. The dBuV metric is also employed to predict proper lubrication intervals. Ultrasound testing is often utilized alongside vibration analysis.

Ultrasound is typically applied in the shock pulse method (SPM) for condition monitoring. This technique uses signals coming from rotating bearings as the baseline for efficient monitoring of machines. For example, imagine a metal ball hitting a metal bar. When the ball contacts the bar, a pressure wave spreads through both materials. The pressure wave is quickly damped out (transient). When the front of the pressure wave hits the shock pulse transducer, it causes a damped-out, back-and-forth movement of the transducer's mass. When the oil film on a bearing is thick, the shock pulse level is low (showing low peaks); when the level increases, the oil film thickness is reduced (Ozturk, 2019).

- **Acoustic Emissions:** ISO 22096 outlines the monitoring of energy through acoustic emissions. Acoustic emissions are an aspect of vibration analysis but occur at much higher frequencies than those detected during traditional vibration analysis. With acoustic emission testing, you are looking for high-frequency signals caused by cracking or impact, as opposed to the repetitive synchronous movement generated by vibration. Acoustic emissions are separate from transducer position, machinery speed, or the configuration of rolling elements (Leaman *et al.*, 2021).

3.8 Condition Monitoring and Predictive Maintenance: Cost-Benefit Tradeoffs

In general, the goal is to maximize the P-F interval (Figure 3.7), which is the time between the first symptoms of impending failure and the functional

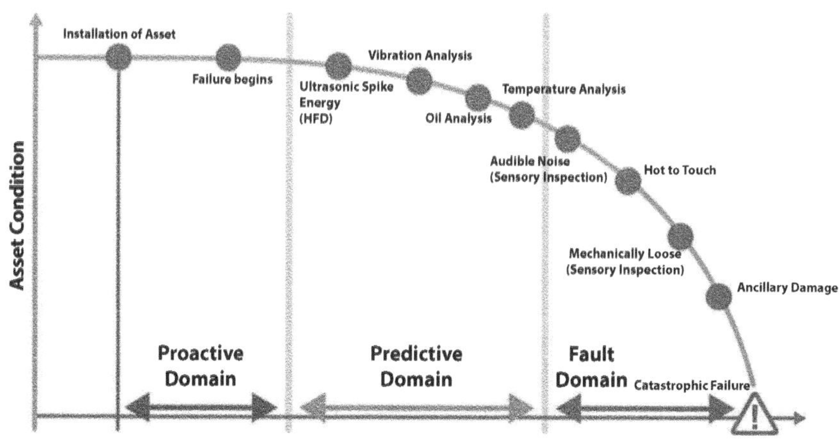

Figure 3.7 P-F diagram.

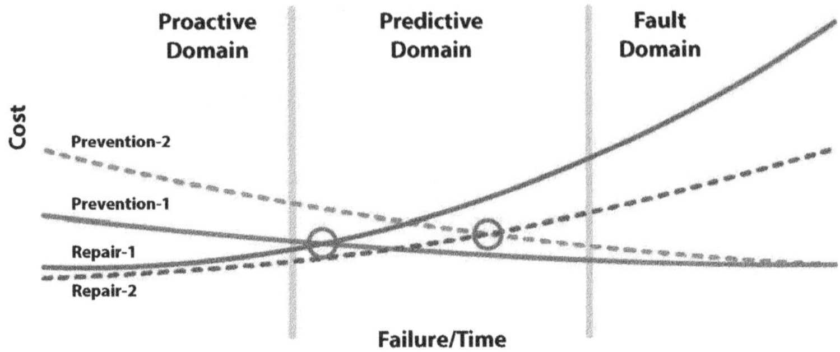

Figure 3.8 Cost-failure/time chart.

failure taking place. In other words, you want to become aware of an impending failure as soon as possible to allow more time for action. This, however, must be balanced with the cost of the methods of prevention, inspection, and detection.

There is a trade-off between the cost of systems to detect and predict the failures and how soon you might detect the condition. Generally, the earlier the detection/prediction, the more expensive it is. However, the longer it takes to detect an impending failure (i.e., the more the asset's condition degrades), the more expensive it is to repair it.

Every asset will have a unique trade-off between the cost of failure prevention (detection/prediction) and the cost of failure. This means some assets probably call for earlier detection methods that come with higher prevention costs like condition monitoring and analytics systems due to the high cost to repair (see the Prevention-1 and Repair-1 curves in the Cost-Failure/Time chart). Some assets may be better suited for more cost-efficient but delayed detection or even a "run-to-failure" model due to lower cost to repair (see the Prevention-2 and Repair-2 curves in the Cost-Failure/Time chart) (Ehrig *et al.*, 2020) (Figure 3.8).

3.9 Conclusion

Condition monitoring plays a pivotal role in ensuring the reliability, safety, and efficiency of equipment and systems in diverse industries. This paper has presented a comprehensive understanding of condition monitoring techniques and their applications, underscoring the significance of proactive maintenance practices. By adopting condition monitoring strategies, organizations can benefit from increased equipment uptime, reduced maintenance costs, improved productivity, and enhanced overall operational efficiency.

The advancements in sensor technologies, data analytics, and machine learning algorithms have further revolutionized the field of condition monitoring. These innovations enable the real-time monitoring of critical parameters, early detection of faults, and the prediction of potential failures. With the integration of the Internet of Things (IoT) and cloud computing, condition monitoring systems can now collect and analyze vast amounts of data, enabling predictive and prescriptive maintenance approaches.

However, it is important to note that effective implementation of condition monitoring requires a holistic approach. This includes establishing clear monitoring objectives, selecting appropriate monitoring techniques, integrating monitoring systems with existing maintenance practices, and ensuring proper data analysis and interpretation. Moreover, organizations must invest in training and development programs to equip their workforce with the necessary skills to operate and maintain condition monitoring systems effectively.

In conclusion, condition monitoring is an indispensable tool for modern maintenance and reliability practices. By continuously monitoring the health of equipment and systems, organizations can achieve significant improvements in operational efficiency, reduce costs, and enhance overall asset performance. As technology continues to advance, the future of condition monitoring holds even greater potential, allowing for more accurate and intelligent maintenance decisions that will drive sustainable and reliable operations.

References

Chavan, N.P., Kale, M.S., Deshmukh, S.P., Thakar, C.M., A Review on Development and Trend of Condition Monitoring and Fault Diagnosis, in: *ECS Transactions*, vol. 107, p. 17863, IOP Publishing, 2022. https://iopscience.iop.org/article/10.1149/10701.17863ecst/meta.

Ehrig, L., Atzberger, D., Hagedorn, B., Klimke, J., Döllner, J., Customizable asymmetric loss functions for machine learning-based predictive maintenance. *2020 8th International Conference on Condition Monitoring and Diagnosis (CMD)*, IEEE, pp. 250–253, 2020.

ElMasry, G., ElGamal, R., Mandour, N., Gou, P., Al-Rejaie, S., Belin, E., Rousseau, D., Emerging thermal imaging techniques for seed quality evaluation: Principles and applications. *Food Res. Int.*, 131, 109025, Elsevier, 2020.

Leaman, F., Baltes, R., Clausen, E., Comparative case studies on ring gear fault diagnosis of planetary gearboxes using vibrations and acoustic emissions. *Forsch. Ingenieurwes.*, 85, 2, 619–628, 2021.

Liu, Z. and Zhang, L., A review of failure modes, condition monitoring and fault diagnosis methods for large-scale wind turbine bearings. *Measurement*, 149, 107002, Elsevier, 2020.

Muthanandan, S. and Nor, K.A.B., Condition monitoring and assessment for rotating machinery, in: *Rotating Machineries: Aspects of Operation and Maintenance*, vol. pp, pp. 1–22, Springer, 2019. https://link.springer.com/chapter/10.1007/978-981-13-2357-7_1.

Nithin, S.K., Hemanth, K., Shamanth, V., Mahale, R.S., Sharath, P.C., Patil, A., Importance of condition monitoring in mechanical domain. *Mater. Today: Proc.*, 54, 234–239, Elsevier, 2022. https://www.sciencedirect.com/science/article/pii/S2214785321057448.

Ozturk, S., *Forecasting Wind Turbine Failures and Associated Costs: Investigating Failure Causes, Effects and Criticalities, Modeling Reliability and Predicting Time-to-Failure, Time-to-Repair and Cost of Failures for Wind Turbines Using Reliability Methods and Machine Learning Techniques*, Columbia University, 2019.

Shafiq, M., Kauhaniemi, K., Robles, G., Isa, M., Kumpulainen, L., Online condition monitoring of MV cable feeders using Rogowski coil sensors for PD measurements. *Electr. Power Syst. Res.*, 167, 150–162, Elsevier, 2019.

Sharp, M., Dadfarnia, M., Sprock, T., Thomas, D., Procedural Guide for System-Level Impact Evaluation of Industrial Artificial Intelligence-Driven Technologies: Application to Risk-Based Investment Analysis for Condition Monitoring Systems in Manufacturing. *J. Manuf. Sci. Eng.*, 144, 7, American Society of Mechanical Engineers Digital Collection, 2022. https://pubmed.ncbi.nlm.nih.gov/36578324/.

Soliman, M.H.A., *Machinery Oil Analysis & Condition Monitoring: A Practical Guide to Sampling and Analyzing Oil to Improve Equipment Reliability*, Mohammed Hamed Ahmed Soliman, 2020.

Tang, X., *Investigation of Data Compression Methods for Intelligent Machine Condition Monitoring*, University of Huddersfield, PhD Thesis, 2020. https://eprints.hud.ac.uk/id/eprint/35227/.

4

Advanced Maintenance Techniques

Davinder Singh* **and Talwinder Singh**

Mechanical Engineering Department, Punjabi University, Patiala, Punjab, India

Abstract

In the previous three decades, the manufacturing industry has seen unparalleled levels of change, including dramatic shifts in management styles, product and process technologies, consumer expectations, supplier attitudes, and competitive behavior. Global competitiveness among organizations has resulted in increasing expectations of industrial organizations in today's highly dynamic and rapidly changing environment. Customers and competitors in the manufacturing and service sectors have increased their pressure on the global marketplace. The data-driven perspective is now the preferred way to increase the quality of maintenance for machines and processes in industrial environments, thanks to advancements in machine learning algorithms and sensor technology. The significance of the purpose of maintenance and the management of maintenance has grown over time. The present study reviews existing literature on maintenance management and highlights the advanced maintenance techniques that will be beneficial for manufacturers and policy makers willing to enhance their manufacturing performance by implementing these techniques.

Keywords: Advanced maintenance techniques, preventive maintenance, total predictive maintenance, condition-based maintenance, intelligent maintenance system, smart maintenance

4.1 Introduction

The importance of the maintenance function, and thus maintenance management, has expanded over time. With greater global competition, the focus has switched from improving efficiency through economies of scale

**Corresponding author*: davinder5206@yhaoo.co.in

Chandan Deep Singh, Davinder Singh, Kanwal Jit Singh and Harleen Kaur (eds.) Condition-Based Maintenance and Residual Life Prediction, (65–86) © 2025 Scrivener Publishing LLC

and internal specialization to meeting market demands for flexibility, delivery performance, and quality [1]. The contemporary business environment is characterized by severe supply-side competition and increased volatility in client expectations on the demand side. These transformations have left indelible imprints on various aspects of manufacturing organizations [2]. To address the difficulties of today's competitive climate, industrial organizations must include quality and performance improvement programs in all elements of their operations. In an increasingly global economy, cost-effective manufacturing has become a necessity to stay competitive [3].

With the introduction of advanced manufacturing and just-in-time (JIT) manufacturing techniques, the nature of manufacturing technology has changed significantly. However, the benefits gained from these applications are often limited by unreliable or inflexible tools [4]. Historically, management has attempted to improve production efficiency by studying, measuring, reporting, and analyzing production costs. A similar effort to improve the efficiency of the maintenance function is long overdue. A general neglect of maintenance as a competitive strategy and a general lack of synergy between maintenance management and quality improvement strategies have been observed in organizations. Therefore, deficiencies in past maintenance practices negatively affect the organization's competitiveness, thereby reducing the power and reliability of production equipment, leading to rapid deterioration of production equipment, and reducing equipment availability due to excessive downtime. The system leads to a decrease in production quality and an increase in inventory. This leads to unreliable delivery performance [5].

4.1.1 Challenge of Maintenance Function

Maintenance is usually considered less efficient than other major budget items. However, most businesses can reduce maintenance costs by at least a third and improve productivity levels by prioritizing maintenance that is essential to management. This priority extends to all levels of an organization's management structure, and each level must understand the importance of maintenance in the success or failure of the organization's goals [2]. Streamline maintenance processes, eliminate waste, and create breakthrough performance in the areas your customers care about. Equipment maintenance is an important component of operating costs for transportation, facilities, mining, and construction. The potential impact of maintenance on production performance is significant. Maintenance is responsible for controlling the cost of labor, materials, tools, and overhead [6].

To thrive in a competitive environment and achieve world-class production, organizations must have both efficient maintenance and effective production strategies. Effective integration of maintenance functions with engineering and other manufacturing functions in an organization saves time, money, and other valuable resources when addressing issues of reliability, availability, serviceability, and performance [3]. Strategic investments in maintenance functions can lead to improved production system performance and can increase the organization's competitive market position. This enables leading organizations worldwide to move beyond traditional firefighting corrective maintenance approaches and adopt effective and efficient protection strategies [7].

The changing needs of the physical assets and equipment over time have been putting tremendous pressure on the maintenance management to adapt proactively to meet the fast-changing requirements of the production systems. Maintenance, being an important support function in businesses with significant investments in plants and machinery, plays an important role in meeting this tall order. Consequently, equipment management has passed through significant changes in recent times. In the present manufacturing scenario, the maintenance function has become an integral part of the overall profitability of an organization [8].

It has been accepted beyond any doubt that maintenance, as a support function in businesses, plays an important role in backing up many emerging business and operation strategies like lean manufacturing, just-in-time production, total quality control, and Six Sigma programs. To that end, the effectiveness of maintenance needs to be improved.

4.2 Traditional Maintenance Techniques

The technical meaning of maintenance involves functional checks, servicing, repairing, or replacing of necessary devices, equipment, machinery, building infrastructure, and supporting utilities in industrial, business, and residential installations. The maintenance techniques have been classified by various researchers in the available literature. These techniques are discussed in detail as follows.

4.2.1 Preventive Maintenance (PM)

Preventive maintenance constitutes a set of tasks performed at a frequency determined by the passage of time, production volume, and machine conditions, aimed at extending the life of an asset or detecting when it is worn out, out of service, or nearing failure. Preventive maintenance (PM) is "a regular inspection aimed at noticing minor problems and correcting them before major problems occur." Ideally, "nothing goes wrong" [9].

The main goal behind PM is to ensure that equipment runs from one scheduled service to the next without failure due to fatigue, neglect, or normal wear and tear (preventable causes). This helps in planned maintenance and condition-based maintenance to be achieved through replacement. Repair worn parts before they actually fail. Maintenance activities include partial or complete overhauls, oil changes, lubrication, minor adjustments, etc., at specified intervals. Additionally, workers can log equipment failures so they know that worn parts need to be replaced or repaired before system failure occurs. Figure 4.1 shows the five main steps of a preventive maintenance program.

The main objectives of PM are:

- Enhance capital equipment productive life.
- Reduce critical equipment breakdown.
- Minimize production loss due to equipment failures.

Preventive maintenance (PM) has the following meanings:

- Personnel whose objective is to maintain equipment in satisfactory operating condition by systematically inspecting,

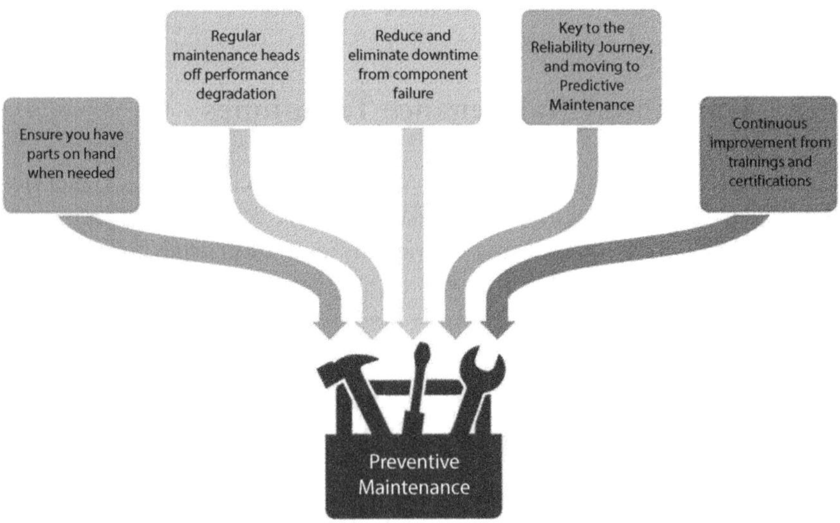

Figure 4.1 Five steps of preventive maintenance plan [9].

identifying, and correcting infant failures before they occur or become major failures.
- Care and maintenance through actions taken to prevent equipment breakdown or failure. These are regular and routine actions performed on equipment to prevent equipment failure.
- Maintenance such as testing, measuring, adjusting, replacing parts, cleaning, etc., which is done specifically to prevent breakdowns from occurring [10].

4.2.2 Condition-Based Maintenance

Condition-based maintenance (CBM), briefly described, is maintenance as needed. Although chronologically much older, it is considered a part or practice within the broader, newer field of predictive maintenance, where new artificial intelligence technologies and connectivity capabilities are being put to practical use. Self-maintenance CBM maintenance is performed after the equipment fails or when one or more indicators indicate that the equipment is performing poorly. This concept can be applied to mission-critical systems that include active redundancy and error reporting. It is also applicable for non-critical systems that lack redundancy and failure reporting [11].

Condition-based maintenance was introduced to maintain the right equipment at the right time. CBM is based on the use of real-time data to prioritize and optimize maintenance resources. Monitoring the state of the system is known as status monitoring. Such systems determine the health of the equipment and operate only when maintenance is actually required. Recent advances have made it possible to accurately instrument equipment, and along with better tools for analyzing condition data, today's maintenance personnel are more aware than ever of when to perform equipment maintenance. Ideally, condition-based maintenance ensures that maintenance personnel do only the right things, minimizing the cost of spare parts, system downtime, and time spent on maintenance.

Advantages of CBM

- Improved system reliability.
- Decreased maintenance costs.
- Decreased maintenance operations, thereby reducing the influence of human error.

Disadvantages of CBM

- High installation costs, for minor equipment items, often more than the value of the equipment items.
- Unpredictable maintenance periods, causing costs to be divided unequally.
- Increased number of parts (the CBM installation itself) that need maintenance and checking.

Today, due to its costs, CBM is not used for less important parts of machinery despite its obvious advantages. However, it can be found everywhere where increased safety is required, and in the future, it will be applied even more widely [12].

4.2.3 Total Productive Maintenance (TPM)

Developed in Japan, TPM focuses on solving maintenance problems through the quality circle method. The benefits of implementing TPM in your organization include a better understanding of equipment performance, better teamwork, and a less adversarial approach between production and maintenance.

Total Productive Maintenance (TPM) began as a physical asset management method that focused on maintaining and improving production machinery to reduce an organization's operating costs. After establishing the PM Award in 1971 and awarding it to Denso Japan, the Japan Institute of Plant Maintenance (JIPM) developed the concept of lean manufacturing to involve all sectors of manufacturing and non-manufacturing industries. It extended the award to eight activities of manufacturing companies. TPMs are designed to decentralize responsibility for machine maintenance and operation and improve employee participation and teamwork in management, engineering, maintenance, and operations [13].

There are eight types of activities in the TPM implementation process:

- Kobetsu-Kaizen (Focused Improvement) Activity
- Jishu-Hozen (Autonomous Maintenance) Activity
- Planned Maintenance Activity
- Hinshitsu-Hozen (Quality Maintenance) Activity
- Development Management Activity
- Education and Training Activity
- OTPM (Office Total Productive Maintenance, or Office TPM)
- Safety, Health, and Environment Activity

The main objective of TPM is to increase the Overall Equipment Effectiveness (OEE) of plant equipment. TPM addresses the causes of accelerated deterioration and production losses while creating the correct environment between operators and equipment to create ownership.

OEE is calculated by multiplying three factors:

Performance x Availability x Quality = OEE

Each factor has two associated losses, making six in total. These six losses are as follows:

Performance = (1) running at reduced speed – (2) Minor Stops
Availability = (3) Breakdowns – (4) Product changeover
Quality = (5) Startup rejects – (6) Running rejects

Finally, the objective is to identify, prioritize, and eliminate the causes of these losses. This is done by self-managing teams that solve problems. Employing consultants to create this culture is a common practice [14].

Principles of TPM:

The eight pillars of TPM are mostly focused on proactive and preventive techniques for improving equipment reliability.

- Autonomous Maintenance – Operators use all of their senses to help identify causes of losses.
- Focused Improvement – A scientific approach to problem-solving to eliminate losses from the factory.
- Planned Maintenance – Professional maintenance activities performed by trained mechanics and engineers.
- Quality Maintenance – A scientific and statistical approach to identifying defects and eliminating their causes.
- Early/Equipment Management – The scientific integration of equipment and design concepts that eliminate losses and facilitate efficient, defect-free production.
- Education and Training – Support to continuous improvement of knowledge of all workers and management
- Safety Health Environmental Conditions
- Administrative and Office TPM – Using total productive maintenance tools to improve all the support aspects of a manufacturing plant including production scheduling, materials management, and information flow, as well as increasing

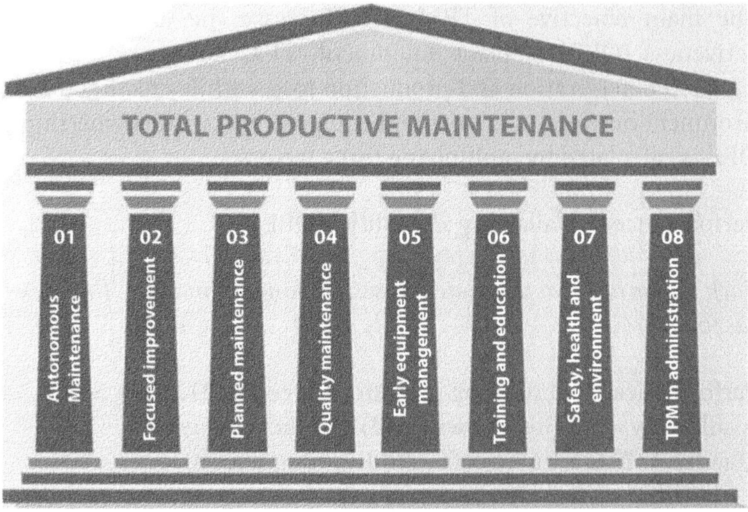

Figure 4.2 Eight pillars of TPM [10].

the morale of individuals and offering awards to well-deserving employees [10].

These eight pillars of TPM are represented in Figure 4.2.

With the help of these pillars, we can increase productivity through improved manufacturing support, such as streamlined operations, better resource management, and the integration of cutting-edge technologies.

4.2.4 Computerized Maintenance Management Systems (CMMS)

A computer maintenance management system (CMMS) is a software package that maintains a computerized database of information about an organization's maintenance activities. This information is used to enable maintenance workers to perform tasks more efficiently (for example, determine which machines need maintenance and which warehouses stock the necessary spare parts). Additionally, administration of CMMS may lead to a better allocation of resources compared to preventive maintenance of each device.

Data from CMMS may also be used to verify regulatory compliance. To properly manage facility maintenance, you need information to analyze what is going on. Doing this manually requires a lot of effort and time. A CMMS also allows you to keep records and track completed and assigned tasks in a timely and cost-effective manner. A CMMS provides the ability to store, retrieve, and analyze information. A total of seven articles have been published in the last eight years [15]. "Front-end maintenance analysis" has been widely used in the industry and has proven to be an effective method for identifying parameters for condition monitoring and clearly showing the relevance of these parameters to maintenance management systems.

Figure 4.3 describes the various steps involved in implementing a CMMS plan.

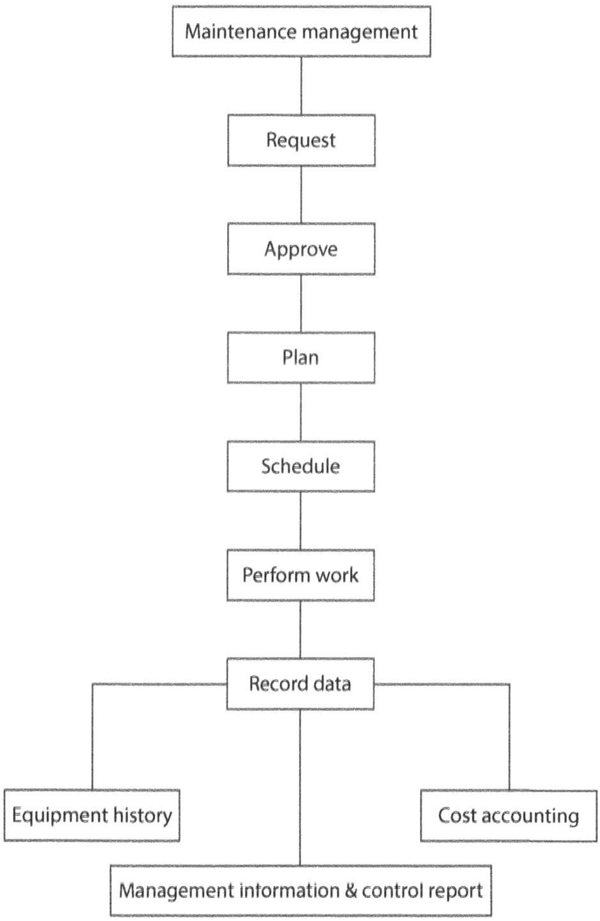

Figure 4.3 Steps in implementing a CMMS plan [15].

4.2.5 Reliability-Centered Maintenance (RCM)

The RCM approach was developed in the 1960s and initially targeted aircraft maintenance. Only in the last decade or so has this concept begun to permeate the industry. It directs maintenance work on parts and units where reliability is critical. A total of five articles have been published in the past few years. Reliability-centered maintenance (RCM) is a maintenance planning concept that ensures that systems continue to do what users want them to do under current operating conditions [16]. Successful implementation of RCM improves cost effectiveness, reliability, device uptime, and provides a better understanding of the level of risk an organization is managing. It is commonly used to achieve improvements in areas such as establishing minimum safe maintenance levels, changing operating procedures and strategies, and establishing capital maintenance regimes and programs [11].

Reliability-centered maintenance is an engineering framework that allows the definition of a complete maintenance program. Maintenance is a means of maintaining the functions that the user may require from the machine under defined operating conditions. It enables mechanical stakeholders to monitor, evaluate, predict, and generally understand the behavior of physical assets. This is included in the first part of the RCM process, which identifies the machine's operating conditions and creates a failure mode, severity, and effects analysis (FMECA). The second part of the analysis is to apply "RCM logic". This helps determine appropriate maintenance tasks for failure modes identified by FMECA. Once the logic of all elements in the FMECA is complete, the resulting maintenance list is "packaged" and the sequence of tasks is simplified for calling into work packages. It is important that the application of maintenance is not compromised at this stage. Lastly, RCM is kept live throughout the "in-service" life of machinery, where the effectiveness of the maintenance is kept under constant review and adjusted in light of the experience gained [7].

The RCM process described in the Department of Defense/United Airlines report recognized three principal risks from equipment failures:

- Threats to safety,
- Threats to operations, and
- Threats to the maintenance budget.

Modern RCM gives threats to the environment a separate classification, though most forms manage them in the same way as threats to safety.

Reliability-centered maintenance (RCM) offers five principal options among the risk management strategies:

- Predictive maintenance tasks,
- Preventive restoration or preventive replacement maintenance tasks,
- Detective maintenance tasks,
- Run-to-failure, and
- One-time changes to the "system" (e.g., changes to hardware design, to operations, or other things) [16].

4.2.6 Predictive Maintenance

Predictive maintenance technology is designed to help determine the condition of in-service equipment to estimate when maintenance should be performed. This approach promises cost savings compared to conventional or time-based preventive maintenance, as tasks are performed only when necessary. Therefore, condition-based maintenance is considered based on an estimate of the item's failure status [6].

The main promise of predictive maintenance is the ability to easily schedule corrective maintenance and prevent unplanned equipment breakdowns. The key aspects are: providing accurate information about the lifespan of equipment, enhancing plant safety, reducing the frequency of accidents that have adverse environmental effects, and optimizing the management and usage of spare parts.

Predictive maintenance differs from preventive maintenance in that it predicts when maintenance will be needed based on the actual condition of the equipment rather than expected life expectancy or life statistics. Machine learning methods are usually used to define the actual state of the system and predict future states [9, 10].

The key components required to implement predictive maintenance include data collection and preprocessing, early failure detection, failure detection, time-to-failure prediction, maintenance scheduling, and resource optimization [17]. Predictive maintenance is also considered one of the drivers of productivity improvement and one of the ways to achieve "just-in-time" in production [14].
Applications (By Industry):

Railway

- Identify warning signs before failure occurs in linear, fixed, and mobile assets.

- Implement a new monitoring system based on the car's cabin to improve safety and detect lane gaps.
- Identify the types of truck assets that have voids and show the severity of the voids.
- Monitor the health of point machines (equipment used to operate railway lines) and help identify early signs of failure before failure.

Manufacturing

- Detection and early detection of defects in production.
- Manufacturers are increasingly collecting big data from Internet of Things (IoT) sensors in factories and products, and using various algorithms on the collected data to detect costly warning signs before detection.

Oil and Gas

- Oil and gas companies often do not have visibility into the condition of their equipment, especially in remote offshore and deep water areas.
- Big data provides oil and gas companies with insights that can analyze and predict equipment failure and the optimal lifetime of systems and components [18].

4.2.7 Risk-Based Maintenance (RBM)

Risk-Based Maintenance (RBM) ensures a sound maintenance strategy that meets the dual goals of minimizing the risks of unplanned equipment breakdowns and a cost-effective strategy. The proposed method is comprehensive and quantitative. It consists of three main modules: risk estimation module, risk assessment module, and maintenance planning module. A case study using this method is also presented.

Risk-based maintenance works on the principle that, in case of failure, maintenance resources are allocated to the most risky assets. For example, if a company has only one full-size semi-truck that needs to be moved every week, that truck will carry a lot of risk if it breaks down [19].

The following are the steps in implementing RBM:

1. **Collect Data About Your Machines and Systems:** The first step is to collect enough data about your current machines

and systems. This allows you to see exactly what is being used and what is happening. This is easily done in UpKeep's Computer Maintenance Management System (CMMS). More on this below.

2. **Failure Risk Assessment:** How likely is it that this device or system will fail? What would cause this system or asset to fail? All of these questions assess the actual risk involved and lead directly to the third step of the process.
3. **Assess the Impact of Failure:** If this device fails, will you be set back a month or two? If this system were to fail, what would be the cost of the outage? These are some of the questions you need to answer when evaluating what exactly will happen if a particular system fails.
4. **Ranking of Assessed Risks and Their Consequences:** The next step in this process is to rank the assessed risks and their consequences in order. Which device or system is most dangerous and which has the least impact when it fails?
5. **Create a Plan:** You now have the information you need to create a risk-based maintenance plan. The first part of this process is to create a time-based inspection schedule. Then develop a plan to mitigate these risks and implement it.
6. **Repeat the cycle regularly:** When performing risk-based maintenance, you should check your company's assets from time to time. Depending on your specific needs, this can be monthly, quarterly, or annually [20].

The important thing to do is to make sure this cycle repeats on a regular basis.

4.2.8 Breakdown Maintenance (BM)

It refers to a maintenance strategy where repairs are carried out after equipment failure/outage or when severe performance degradation occurs [4]. Before 1950, this maintenance strategy was used primarily by manufacturing organizations around the world. At this stage, device repairs are performed only when urgently needed. This concept suffers from unplanned outages, excessive damage, spare parts problems, high repair costs, excessive waiting and maintenance time, and troubleshooting problems [6].

4.3 Advanced Maintenance Techniques

4.3.1 Intelligent Maintenance System (IMS)

Intelligent Maintenance System (IMS) is a system that uses data collected from machines to predict and prevent possible machine breakdowns. Car breakdowns can be costly and, in some cases, catastrophic. To prevent breakdowns, a system is needed that analyzes the machine's behavior and issues warnings and preventive maintenance instructions. Analysis of machine behavior is possible with advanced sensors, data acquisition systems, data storage/transfer capabilities, and data analysis tools. These are the same set of tools developed for forecasting. A set of data collection, storage, transformation, analysis, and decision-making for intelligent maintenance is called Intelligent Maintenance System (IMS) [21].

Intelligent maintenance systems are systems that use data analysis and decision support tools to predict and prevent possible machine breakdowns. Recent advances in information technology, computers, and electronics have facilitated the design and implementation of such systems.

The key research elements of intelligent maintenance systems consist of:

- Transforming data into information and knowledge, and synchronizing decision-making with remote systems
- Intelligent built-in predictive algorithms to assess degradation and predict future performance
- Software and hardware platform for running online models
- Embedded product services and lifecycle information for closed-loop product design.

E-Manufacturing and E-Maintenance:

As the use of tetherless communication technologies (such as the Internet) evolves, e-intelligence is having a major impact on industries. These impacts are causing companies to shift their manufacturing operations from traditional factory integration practices to e-factory and e-supply chain philosophies. These changes transform companies from local factory automation to global business automation. The purpose of electronic manufacturing is to predict the potential for product quality change and equipment loss from factory floor assets. It provides predictive capabilities for machine maintenance [22].

Figure 4.4 represents the overview of an IMS.

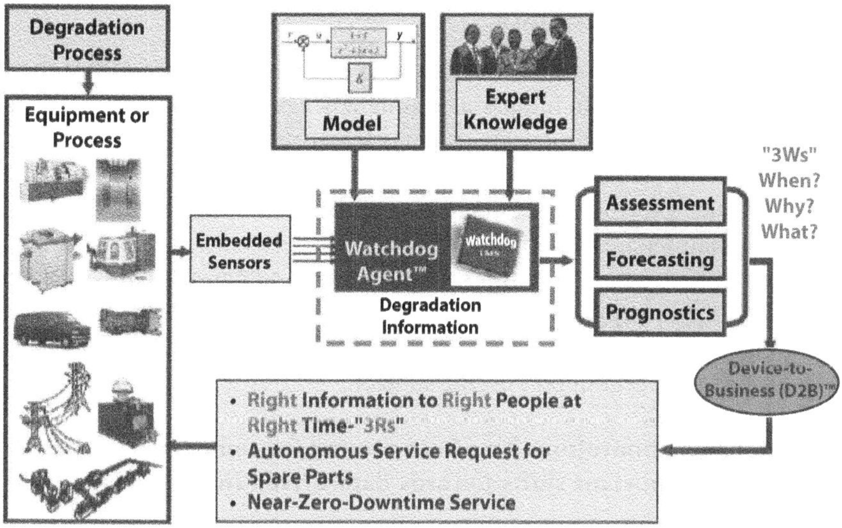

Figure 4.4 Overview of an intelligent maintenance system [21].

The major functions and objectives of e-manufacturing are:

- Provide a transparent, seamless, and automated information exchange process to enable a once-for-ever (Ohio) environment.
- Improve asset utilization on the plant floor using a holistic approach that incorporates predictive maintenance technology tools.
- Integrating supply chain management (SCM) operations with asset optimization while employing advanced predictive intelligence techniques to enhance customer service.

The e-maintenance infrastructure consists of several information sectors:

- Control systems and production schedulers
- Engineering product data management systems
- Enterprise resource planning (ERP) systems
- Condition monitoring systems
- Maintenance scheduling (CMMS/EAM) systems
- Plant asset management (PAM) systems

4.3.2 Smart Maintenance

Smart maintenance is difficult to define. It is a common term rather than an official term found in industrial maintenance manuals. However, its use is becoming more common. At its core, smart maintenance means using tools and technologies that consistently provide the right data at the right time to make informed asset management decisions and increase the efficiency of maintenance operations [23].

Smart maintenance is about using new technologies such as mobile solutions, big data applications, and the Internet of Things to ensure that all equipment required for production is always working at 100% efficiency. Here are some recent quotes on the subject:

"Maintenance innovations have been around for some time, but we are now seeing important shifts towards data-driven innovations and the power of IoT."

In today's competitive market, you need to deliver quality products quickly and affordably. A failure is a disaster, as is a machine or equipment failure that impedes product quality. Therefore, good maintenance is important. Innovations such as big data applications, the Internet of Things, machine learning, and augmented reality are making maintenance smarter [24].

The real power of smart maintenance is the ability to analyze and understand large amounts of raw maintenance data. Imagine having a CMMS that can do everything a modern CMMS system can do, with one exception: it cannot generate reports. To get the desired numbers, you have to open each asset, part, vendor, work order, etc., manually [22].

Often grouped under the Industrial Internet of Things (IIoT) or Industry 4.0 umbrella, companies can generate and analyze vast amounts of maintenance data using a variety of software, sensors, tools, and algorithms.

Advanced algorithms can use those data sets for a number of advanced functions such as:

- Automating specific tasks
- Predicting equipment failures or malfunctions
- Providing personalized recommendations to production and maintenance teams based on real-time production and asset condition data

Smart maintenance usually involves a combination of various technologies. Figure 4.5 describes various technologies that are used in implementing smart maintenance.

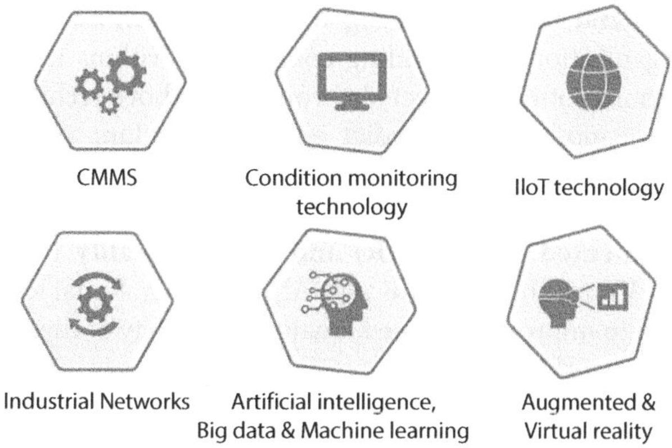

Figure 4.5 Technologies used in implementing smart maintenance [24].

Following is a brief description of these technologies:

1. **Computer Maintenance Management System (CMMS):** Used to capture, store, access, and manage maintenance data and automate maintenance workflows such as scheduling and reporting.
2. **Condition Monitoring Technology:** Raw data from condition monitoring sensors are used in advanced maintenance strategies such as predictive maintenance (PdM) and reliability-centered maintenance (RxM). This includes temperature, vibration, and noise sensors, as well as techniques such as infrared thermography, ultrasonic inspection, corrosion monitoring, and motor current analysis.
3. **Industrial Internet of Things (IIoT) Technologies:** Proximity sensors, pressure sensors, smoke sensors, motion sensors, image sensors, humidity sensors, and light sensors are just a few types of interconnected sensors that the Internet of Things is using to collect real-time data about machines and facilities. Uses. These help eliminate inefficiencies, improve quality control, and create a more sustainable work environment.
4. **Industrial Networks:** Industrial networks can manage a large amount of data exchanged between several devices and in large spaces. It helps connect sensor and machine data with industrial applications such as manufacturing systems, CMMS software, and energy management applications.

5. **Artificial Intelligence, Big Data, and Machine Learning:** Organizations are building complex algorithms using vast amounts of data collected through the above technologies. These models help predict equipment failure, implement just-in-time parts inventory management, optimize production, and reduce downtime and overall maintenance costs.
6. **Augmented Reality (AR) and Virtual Reality (VR):** AR and VR technologies are applied at the operational level and used to improve the speed, quality, and safety of the maintenance work performed.

This is by no means an exhaustive list. You also don't need all these tools or a complete digital transformation to effectively implement a smart maintenance program. Smart maintenance can encompass a wide range of capabilities and technologies to meet the unique needs of your organization [25].

The Benefits of Smart Maintenance:

1. **Increased Productivity:** The company will be more productive because your machines are always on. Service technicians are also more efficient thanks to insight into how to maintain or repair equipment.
2. **Lower Maintenance Costs:** Improper maintenance timing leads to breakdowns and high costs. The right balance of preventive and predictive maintenance can reduce parts and labor costs.
3. **Extend Equipment Life:** Having a clear understanding of machine/equipment conditions and performing scheduled maintenance will ensure that your equipment lasts longer.
4. **Ensure Compliance:** Reporting capabilities and procedures enabled by smart maintenance solutions make it easy to comply with all maintenance-related standards, such as ISO 55000 standards.
5. **Increased Safety, Reduced Energy Consumption, and CO_2 Emissions:** Timely monitoring and correction of potential problems enable employees to work in safer conditions. In addition, advanced tools can help reduce energy consumption while reducing CO_2 emissions [24].

4.4 Conclusions

This chapter focuses on the contribution of various advanced maintenance technologies (AMTs) to achieve strategic gains to face the challenges posed by global competition [23]. These techniques have been proposed as the main competitive strategies of corporate organizations in the global market. An effective implementation program can focus on addressing the organization's maintenance issues with the goal of optimizing equipment performance. Recently, intelligent maintenance has become the new management paradigm for organizations of all types, and many have shown that this maintenance approach can achieve significant business improvements. Maintenance concepts and philosophies can be effectively applied to make fundamental improvements in an organization's production performance, thereby preparing the organization for success in a highly competitive environment. AMT has proven to be an effective global strategy for the continuous improvement of company performance in terms of achieving strategic core competencies. Therefore, in a competitive scenario, AMT is one of the best proactive strategic initiatives that can lead organizations to new levels of scalable performance and truly separate organizational success from failure.

References

1. Ahmad, M.M. and Dhafr, N., Establishing and improving manufacturing performance measures. *Rob. Comput. Integr. Manuf.*, 18, 3-4, 171–176, 2002.
2. Smith, R., Key Performance Indicators-Leading or lagging and when to use them, in: *Smith and Hawkins Lean Maintenance: Reduce Costs, Improve Quality, and Increase Market Share*, pp. 249–252, Elsevier BV, Burlington, 2004.
3. Brundage, M.P., Morris, K.C., Sexton, T., Moccozet, S., Hoffman, M., Developing maintenance key performance indicators from maintenance work order data. *Paper Presented at the ASME 2018 13th International Manufacturing Science and Engineering Conference*, MSEC, 2018.
4. Smith, R. and Mobley, R.K., *Rules of Thumb for Maintenance and Reliability Engineers*, Butterworth-Heinemann, Oxford, UK, 2011.
5. Akkermans, H., Besselink, L., Van Dongen, L., Schouten, R., Smart moves for smart maintenance, 2016, available at: https://pdfs.semanticscholar.org/2b-d7/53952261520bdccbeb026675245a77ca400a.pdf (accessed 1 October 2018).

6. Hipkin, I.B. and Cock, C.D., TQM and BPR: lessons for maintenance management. *Omega – The International Journal of Management Science*, vol. 28, pp. 277–92, 2000.
7. Seth, D. and Tripathi, D., Relationship between TQM and TPM implementation factors and business performance of manufacturing industry in Indian context. *Int. J. Qual. Reliab. Manage.*, 22, 2/3, 256–77, 2005.
8. Ahuja, I.P.S., Khamba, J.S., Choudhary, R., Improved organizational behavior through strategic total productive maintenance implementation. *Paper No. IMECE2006-15783, ASME International Mechanical Engineering Congress and Exposition (IMECE)*, pp. 1–8, November 5-10, 2006.
9. Ahuja, I.P.S., Singh, T.P., Sushil, M., Wadood, A., Total productive maintenance implementation at Tata Steel for achieving core competitiveness. *Productivity*, 45, 3, 422–6, 2004.
10. Thun, J.H., Maintaining preventive maintenance and maintenance prevention analyzing the dynamic implications of total productive maintenance. *Syst. Dyn. Rev.*, 22, 2, 163–79, 2006.
11. Tripathi, D., Influence of experience and collaboration on effectiveness of quality management practices: the case of Indian manufacturing. *Int. J. Prod. Perform. Manage.*, 54, 1, 23–33, 2005.
12. Shamsuddin, A., Hassan, M.H., Taha, Z., TPM can go beyond maintenance: excerpt from a case implementation. *J. Qual. Maint. Eng.*, 11, 1, 19–42, 2005.
13. Holmgren, M., Maintenance-related losses at the Swedish Rail. *J. Qual. Maint. Eng.*, 11, 1, 5–18, 2005.
14. Wijesinghe, D. and Mallawarachchi, H., A systematic approach for maintenance performance measurement: apparel industry in Sri Lanka. *J. Qual. Maint. Eng.*, 25, 1, 41–53, 2019.
15. Dalenogare, L.S., Benitez, G.B., Ayala, N.F., Frank, A.G., The expected contribution of Industry 4.0 technologies for industrial performance. *Int. J. Prod. Econ.*, 204, 383–394, 2018.
16. Ante, G., Facchini, F., Mossa, G., Digiesi, S., Developing a key performance indicators tree for lean and smart production systems. *IFAC-PapersOnLine*, 51, 11, 13–18, 2018.
17. Parida, A., Kumar, U., Galar, D., Stenström, C., Performance measurement and management for maintenance: a literature review. *J. Qual. Maint. Eng.*, 21, 1, 2–33, 2015.
18. Alsyouf, I., Measuring maintenance performance using a balanced scorecard approach. *J. Qual. Maint. Eng.*, 12, 2, 133–49, 2006.
19. Ireland, F. and Dale, B.G., A study of total productive maintenance implementation. *J. Qual. Maint. Eng.*, 7, 3, 183–91, 2001.
20. Jeong, K.-Y. and Phillips, D.T., Operational efficiency and effectiveness measurement. *Int. J. Oper. Prod. Manage.*, 21, 11, 1404–16, 2001.
21. McKone, K.E., Roger, G.S., Cua, K.O., The impact of total productive maintenance practices on manufacturing performance. *J. Oper. Manage.*, 19, 1, 39–58, 2001.

22. Tirabeni, L., De Bernardi, P., Forliano, C., Franco, M., How can organisations and business models lead to a more sustainable society? A framework from a systematic review of the industry 4.0. *Sustainability*, 11, 22, 6363, 2019.
23. Bokrantz, J., Skoogh, A., Berlin, C., Wuest, T., Stahre, J., Smart maintenance: an empirically grounded conceptualization. *Int. J. Prod. Econ.*, 223, 1–17, 2020.
24. Lundgren, C., Bokrantz, J., Skoogh, A., Performance indicators for measuring the effects of Smart Maintenance. *Int. J. Prod. Perform. Manage.*, 70, 6, 1291–1316, 2021.
25. Bokrantz, J., Skoogh, A., Berlin, C., Wuest, T., Stahre, J., Smart Maintenance: a research agenda for industrial maintenance management. *Int. J. Prod. Econ.*, 224, 1–14, 2020.

5

Unveiling the Future: Residual Life Prediction for Enhanced Asset Management

Maninder Singh*, Mukhtiar Singh, Jasvinder Singh, Mandeep Singh and Harjit Singh

School of Mechanical Engineering, Lovely Professional University, Phagwara, Punjab, India

Abstract

This book chapter delves into the fascinating field of residual life prediction, an emerging discipline within asset management that aims to forecast the remaining useful life of critical components and systems. As industries worldwide face the challenges of aging infrastructure and increasing operational costs, accurate prediction of the residual life of assets has become a crucial aspect of maintenance planning, cost optimization, and risk mitigation. The chapter begins by providing an overview of the concept of residual life prediction and its significance in various industries, including manufacturing, energy, transportation, and healthcare. It explores the theoretical foundations and methodologies used to predict the remaining life of assets, encompassing both empirical models and advanced data driven techniques. Key factors influencing residual life, such as environmental conditions, usage patterns, and maintenance history, are also investigated. Furthermore, the chapter examines the role of advanced technologies in enhancing residual life prediction, including machine learning, sensor networks, and Internet of Things (IoT) applications. It explores how these technologies enable the collection and analysis of vast amounts of data, enabling more accurate and proactive predictions.

Practical implementation challenges and considerations are discussed, encompassing data acquisition, quality assurance, model selection, and integration into existing asset management systems. The chapter also explores the importance of continuously updating and refining prediction models as new data becomes

**Corresponding author*: maninder.17902@lpu.co.in

Chandan Deep Singh, Davinder Singh, Kanwal Jit Singh and Harleen Kaur (eds.) Condition-Based Maintenance and Residual Life Prediction, (87–100) © 2025 Scrivener Publishing LLC

available, ensuring their ongoing accuracy and effectiveness. To illustrate the real-world applications of residual life prediction, several case studies across different industries are presented. These case studies highlight the benefits of accurate predictions in terms of reducing downtime, optimizing maintenance strategies, extending asset lifespan, and ultimately improving operational efficiency and profitability. In conclusion, this book chapter emphasizes the significance of residual life prediction in contemporary asset management practices. By harnessing the power of data analytics and advanced modeling techniques, organizations can optimize their maintenance strategies, mitigate risks, and make informed decisions regarding asset replacement or rejuvenation. Residual life prediction is poised to revolutionize asset management and enable a more sustainable and cost-effective future for industries worldwide.

Keywords: Residual life prediction, asset management, condition-based maintenance, data-driven techniques, machine learning, Internet of Things (IoT), operational efficiency, cost optimization

5.1 Introduction

Residual life prediction is a vital aspect of asset management, enabling organizations to proactively plan maintenance activities, optimize resource allocation, and mitigate risks associated with aging infrastructure. Accurate estimation of the remaining useful life of critical components and systems has gained significant attention in recent years, as it plays a crucial role in maximizing operational efficiency, reducing downtime, and minimizing costs. This book chapter aims to provide a comprehensive overview of residual life prediction, its significance, methodologies, and practical implications. One of the key aspects of residual life prediction is understanding the economic impact of asset failures and the benefits of proactive maintenance planning. Several studies highlight the substantial financial losses incurred due to unplanned downtime and reactive maintenance practices (Wu *et al.*, 2018). By accurately predicting the residual life of assets, organizations can implement condition-based maintenance strategies, optimize the timing of replacements, and allocate resources more effectively, resulting in improved operational performance and cost savings (Ivanov *et al.*, 2021).

However, accurate residual life prediction faces various challenges and limitations. Factors such as uncertainties in degradation processes, variations in operational conditions, limited historical data, and the complexity of interactions between multiple components and subsystems pose significant challenges in developing reliable prediction models (Wang *et al.*, 2020).

Overcoming these challenges is essential to enhance the accuracy and effectiveness of residual life prediction in real-world scenarios. To address these challenges, researchers and practitioners have developed various methodologies and techniques for residual life prediction. Traditional approaches such as reliability-based models, physics-based models, and statistical techniques have been widely used (Jardine *et al.*, 2006). Additionally, advanced data-driven techniques, including machine learning algorithms and artificial intelligence, have shown promising results in improving the accuracy of residual life predictions (Rahmani *et al.*, 2021). Furthermore, the integration of advanced technologies such as sensor networks, the Internet of Things (IoT), and big data analytics has revolutionized residual life prediction. These technologies enable real-time data collection, monitoring, and analysis, allowing for more accurate and timely predictions of asset health and remaining life (Luo *et al.*, 2021). The availability of large-scale data sets combined with advanced analytics algorithms enhances the understanding of asset degradation patterns and facilitates proactive decision-making. To showcase the practical applications and effectiveness of residual life prediction, this chapter includes relevant case studies from various industries. These case studies highlight how accurate predictions have been applied to optimize maintenance strategies, extend asset lifespan, and minimize operational disruptions. Examples from sectors such as manufacturing, energy, transportation, and healthcare demonstrate the wide-ranging benefits of residual life prediction in diverse real-world scenarios (Nguyen *et al.*, 2019; Rashid *et al.*, 2024).

5.1.1 Overview of the Key Challenges and Limitations in Accurate Estimation

Accurately estimating the residual life of assets poses several challenges and limitations due to the complex nature of degradation processes, variations in operational conditions, data availability, and the need for comprehensive models. Understanding and addressing these challenges are crucial for improving the accuracy and reliability of residual life predictions. This section provides an overview of the key challenges and limitations in the accurate estimation of residual life, supported by relevant references.

Uncertainties in Degradation Processes: Asset degradation is influenced by multiple factors, including usage patterns, environmental conditions, and maintenance practices. However, capturing the complex degradation mechanisms and accurately modeling their interactions pose significant challenges (Li *et al.*, 2019). The uncertainties associated with degradation processes make it difficult to predict the remaining life of assets with high precision.

Limited Availability of Historical Data: Accurate residual life prediction often relies on historical data that captures the performance and condition of assets over time. However, in many cases, the availability of reliable and sufficient historical data is limited, particularly for assets with long operational lifespans or assets in emerging industries (Zhang *et al.*, 2020). Insufficient data can impede the development of accurate prediction models and hinder their practical implementation.

Variations in Operational Conditions: Asset performance and degradation can vary significantly due to changes in operational conditions, such as load profiles, environmental factors, and maintenance activities. Incorporating these variations into prediction models presents challenges as the relationships between degradation and operational conditions are often complex and non-linear (Xu *et al.*, 2021). Failure to account for these variations can lead to inaccurate residual life estimates.

Complex Interactions Between Components and Subsystems: Assets are composed of multiple interconnected components and subsystems. The interactions between these elements can influence the degradation patterns and overall asset performance. Modeling and quantifying these complex interactions require a comprehensive understanding of the asset's structure, functions, and dependencies (Chen *et al.*, 2018). The complexity of these interactions adds to the challenges in accurately estimating residual life.

Data Quality and Reliability: The accuracy of residual life prediction heavily relies on the quality and reliability of the data used for model development. Data issues, such as missing values, outliers, and measurement errors, can introduce biases and uncertainties into the prediction models, affecting their accuracy and reliability (Vachtsevanos *et al.*, 2017). Ensuring data quality and implementing appropriate data preprocessing techniques are essential for accurate residual life estimation.

Limited Understanding of Rare Events: Rare events, such as extreme operating conditions or unforeseen failures, can have a significant impact on asset degradation and remaining life. However, accurately capturing and modeling these rare events is challenging due to their infrequent occurrences and limited data availability (Borges *et al.*, 2020). Incorporating rare events into prediction models is crucial for capturing the full range of potential failure scenarios. Addressing these challenges requires ongoing research efforts and the development of advanced modeling techniques that consider the complexities and uncertainties associated with asset degradation. By mitigating these limitations, more accurate and reliable residual life predictions can be achieved, enabling organizations to optimize maintenance strategies, reduce costs, and improve asset management practices.

5.1.2 Objectives of the Chapter

- Provide an In-Depth Understanding of Residual Life Prediction: The chapter aims to provide a comprehensive overview of residual life prediction, including its definition, significance, and relevance in the field of asset management. It establishes a solid foundation of knowledge for readers to grasp the fundamental concepts and principles underlying residual life prediction.
- Identify Key Methodologies and Techniques: The chapter aims to explore various methodologies and techniques employed in residual life prediction. It discusses traditional approaches, such as reliability-based models and physics-based models, as well as advanced data-driven techniques like machine learning algorithms and artificial intelligence. By presenting a range of approaches, the chapter enables readers to understand the strengths, limitations, and applicability of different methods.
- Highlight Challenges and Limitations: Understanding the challenges and limitations in accurate estimation of residual life is crucial for developing effective prediction models. The chapter identifies and discusses key challenges such as uncertainties in degradation processes, data availability, variations in operational conditions, and complex interactions between components and subsystems. By highlighting these challenges, the chapter aims to foster a comprehensive understanding of the complexities involved in residual life prediction.
- Showcase Practical Implications and Applications: The chapter presents real-world case studies and examples to demonstrate the practical implications and applications of residual life prediction. These case studies cover various industries, highlighting how accurate predictions can optimize maintenance strategies, improve asset performance, and reduce costs. By showcasing practical examples, the chapter aims to provide readers with insights into the real-world impact and benefits of residual life prediction.

- Discuss Emerging Trends and Future Prospects: The chapter explores emerging trends and future directions in residual life prediction. It discusses advancements in technology, such as sensor networks, the Internet of Things (IoT), and big data analytics, that are shaping the field. By discussing the potential advancements and future prospects, the chapter provides readers with a glimpse into the exciting developments and opportunities in residual life prediction.
- Provide Recommendations and Guidelines: The chapter aims to provide practical recommendations and guidelines for implementing residual life prediction models effectively. It covers aspects such as data acquisition, quality assurance, model development, and integration with existing maintenance management systems. By offering practical guidance, the chapter helps readers navigate the challenges and complexities of implementing residual life prediction in real-world scenarios.
- By addressing these objectives, the chapter aims to equip readers with the knowledge, insights, and practical guidance necessary to leverage residual life prediction for improved asset management practices, optimized maintenance strategies, and enhanced operational efficiency.

5.2 Residual Life Prediction Techniques

5.2.1 Prognostic Models

Prognostic models play a crucial role in residual life prediction by forecasting the remaining useful life of assets based on their current condition and degradation patterns. These models utilize historical data, sensor measurements, and advanced analytical techniques to estimate the future deterioration and remaining lifespan of assets. This section provides an overview of some commonly used prognostic models for residual life prediction, supported by relevant references.

Reliability-Based Models: Reliability-based models use statistical techniques to estimate the residual life of assets based on their observed degradation patterns and historical failure data. These models typically rely on probabilistic approaches, such as survival analysis, Weibull analysis, or proportional hazard models, to quantify the reliability and failure probabilities of assets over time (Nguyen *et al.*, 2018). Reliability-based models are widely used in various industries for residual life prediction.

Physics-Based Models: Physics-based models utilize the understanding of the underlying physical processes and mechanisms causing asset degradation. These models incorporate physics-based equations and principles to simulate the degradation behavior and predict the residual life of assets. Physics-based models often require detailed knowledge of the asset's structure, materials, operating conditions, and degradation mechanisms (Chen *et al.*, 2018). They are particularly useful when the degradation processes are well understood and can be accurately modeled.

Data-Driven Models: Data-driven models leverage machine learning algorithms, artificial intelligence, and data analytics techniques to predict the residual life of assets based on historical data and sensor measurements. These models learn patterns and relationships from the data and use them to make predictions about future asset degradation. Data-driven models, such as support vector machines, random forests, neural networks, and deep learning models, have shown promising results in various applications of residual life prediction (Wang *et al.*, 2021).

Hybrid Models: Hybrid models combine the strengths of multiple approaches, such as reliability-based models and data-driven models, to enhance the accuracy of residual life prediction. These models integrate physics-based understanding with data-driven algorithms to capture both the underlying degradation mechanisms and the complex relationships within the data. Hybrid models aim to leverage the complementary advantages of different modeling approaches and provide more accurate and robust predictions (Ma *et al.*, 2020).

Ensemble Models: Ensemble models combine multiple individual models to generate a collective prediction. These models can include a combination of different prognostic models, data sources, or feature selection techniques. By aggregating the predictions from multiple models, ensemble models aim to improve prediction accuracy, reduce bias, and enhance robustness (Zhang *et al.*, 2019). Ensemble models are particularly useful when dealing with uncertainty and variability in the data or when different models perform well under different conditions.

5.2.2 Statistical Approaches for Residual Life Prediction

Statistical approaches play a significant role in residual life prediction by utilizing historical data, probabilistic models, and statistical techniques to estimate the remaining useful life of assets. These approaches provide a quantitative understanding of asset degradation and enable accurate predictions based on statistical analysis. This section provides an overview of some commonly used statistical approaches for residual life prediction, supported by relevant references.

Survival Analysis: Survival analysis, also known as reliability analysis or time-to-event analysis, is a statistical approach widely used for residual life prediction. It models the time until an event of interest (e.g., failure or degradation exceeding a threshold) occurs. Survival analysis methods, such as Kaplan-Meier estimation, Cox proportional hazards model, and parametric survival models (e.g., Weibull, log-normal), are commonly employed to estimate the remaining useful life of assets based on observed degradation patterns (Cai *et al.*, 2018).

Bayesian Statistics: Bayesian statistics provides a framework for updating beliefs and making predictions based on both prior knowledge and observed data. Bayesian approaches for residual life prediction incorporate prior information about the degradation process, expert opinions, and historical data to estimate the remaining useful life of assets. These models account for uncertainties and provide probabilistic predictions, allowing for better decision-making and risk assessment (Li *et al.*, 2019).

Degradation Modeling: Degradation modeling is a statistical approach that focuses on capturing the degradation patterns of assets over time. It involves modeling the relationship between degradation measurements and the remaining useful life of assets. Various statistical models, such as linear regression, polynomial regression, mixed-effects models, and state-space models, are employed to estimate the remaining life based on degradation data (Feng *et al.*, 2020).

Time Series Analysis: Time series analysis is a statistical approach that examines the sequential dependence and patterns in data collected over time. In the context of residual life prediction, time series models, such as autoregressive integrated moving average (ARIMA), seasonal decomposition of time series (STL), and exponential smoothing methods, can be used to forecast future degradation trajectories and estimate the remaining useful life of assets (Huang *et al.*, 2020).

Monte Carlo Simulation: Monte Carlo simulation is a statistical technique that involves generating multiple random samples based on specified distributions to model uncertainty and estimate outcomes. In residual life prediction, Monte Carlo simulation can be used to simulate the degradation process, assess the variability in the remaining useful life estimates, and quantify the associated uncertainties (Chen *et al.*, 2021).

5.2.3 Machine Learning Techniques for Residual Life Prediction

Machine learning techniques have gained significant attention in the field of residual life prediction due to their ability to learn patterns and relationships from data without explicit programming. These techniques leverage advanced algorithms and data analytics to make accurate predictions about the remaining useful life of assets. This section provides an overview of some machine learning techniques commonly used for residual life prediction, supported by relevant references.

Support Vector Machines (SVM): Support Vector Machines are supervised learning models used for classification and regression tasks. SVM algorithms create hyperplanes in a high-dimensional space to separate data points and make predictions. In residual life prediction, SVM can be employed to model the relationship between degradation features and the remaining useful life of assets, enabling accurate predictions (Wang *et al.*, 2016).

Random Forest: Random Forest is an ensemble learning technique that combines multiple decision trees to make predictions. It is particularly effective in handling complex datasets and capturing non-linear relationships. In residual life prediction, Random Forest algorithms can utilize historical data and sensor measurements to estimate the remaining useful life of assets (Javed *et al.*, 2020).

Neural Networks: Neural networks are computational models inspired by the structure and function of the human brain. They consist of interconnected nodes, or neurons, organized into layers. Neural networks can learn complex patterns and relationships in data, making them suitable for residual life prediction. Deep neural networks, such as Convolutional Neural Networks (CNN) and Recurrent Neural Networks (RNN), have shown promising results in predicting the remaining useful life of assets (Wang *et al.*, 2018).

Gradient Boosting Machines (GBM): Gradient Boosting Machines are ensemble learning algorithms that sequentially combine weak models to create a strong predictive model. GBM algorithms, such as XGBoost and LightGBM, are widely used for residual life prediction tasks. They handle high-dimensional data, capture complex interactions, and provide accurate predictions by iteratively improving the model's performance (Li *et al.*, 2019).

Long Short-Term Memory (LSTM): LSTM is a type of recurrent neural network that can model temporal dependencies in sequential data. LSTM networks are particularly useful for residual life prediction when historical sensor data is available. They can effectively capture long-term dependencies and dynamic patterns, enabling accurate predictions of the remaining useful life (Afridi *et al.*, 2021).

5.3 Applications of Residual Life Prediction

Residual life prediction has diverse applications across various industries and sectors. By estimating the remaining useful life of assets, organizations can optimize maintenance strategies, improve asset performance, reduce downtime, and enhance overall operational efficiency. This section provides an overview of some key applications of residual life prediction, supported by relevant references.

Predictive Maintenance: One of the primary applications of residual life prediction is in predictive maintenance. By accurately estimating the remaining useful life of assets, organizations can schedule maintenance activities proactively, replacing or repairing components before failure occurs. This approach helps prevent unexpected breakdowns, minimize downtime, and reduce maintenance costs (Wang *et al.*, 2018).
Asset Management: Residual life prediction plays a vital role in effective asset management. By knowing the remaining useful life of assets, organizations can optimize asset utilization, plan for asset replacements or upgrades, and allocate resources efficiently. It enables organizations to make informed decisions regarding asset retirement, refurbishment, or replacement (Shi *et al.*, 2019).
Warranty Management: Residual life prediction is valuable for warranty management, especially for manufacturers. By accurately estimating the remaining useful life of products under warranty, manufacturers can proactively address potential issues, plan for warranty coverage, and allocate resources effectively. It helps manufacturers minimize warranty claims, improve customer satisfaction, and optimize warranty policies (Ding *et al.*, 2020).
Risk Assessment and Insurance: Residual life prediction is utilized in risk assessment and insurance industries to evaluate the potential risks associated with assets and determine insurance coverage. Accurate estimation of the remaining useful life helps insurers assess the likelihood of failure or damage and set appropriate premiums. It enables risk managers to make informed decisions regarding risk mitigation and coverage (Bui *et al.*, 2021).

Energy Systems and Infrastructure: Residual life prediction is relevant in energy systems and infrastructure sectors. By estimating the remaining useful life of components in power plants, renewable energy systems, and critical infrastructure, organizations can optimize maintenance schedules, plan for component replacements, and ensure reliable and efficient operations. It helps in reducing downtime, improving energy production, and enhancing the overall performance of energy systems (Khan *et al.*, 2019). Transportation and Fleet Management: Residual life prediction is applicable in transportation and fleet management industries. By estimating the remaining useful life of vehicles, aircraft, or other transportation assets, organizations can optimize maintenance schedules, plan for fleet replacements, and ensure the safety and reliability of transportation operations. It helps in minimizing disruptions, optimizing asset utilization, and reducing maintenance costs (Song *et al.*, 2020).

5.4 Conclusion

Residual life prediction is a critical aspect of asset management and maintenance practices. This chapter has provided an overview of residual life prediction, highlighting its importance and relevance in various industries. We explored the key challenges and limitations faced in accurately estimating residual life, including uncertainties in degradation processes, data availability, and variations in operational conditions.

The objectives of this chapter were to provide readers with an in-depth understanding of residual life prediction, identify key methodologies and techniques, highlight challenges and limitations, showcase practical implications and applications, discuss emerging trends and future prospects, and provide recommendations and guidelines for effective implementation.

We discussed different prognostic models for residual life prediction, including reliability-based models, physics-based models, data-driven models, hybrid models, and ensemble models. These models leverage statistical approaches, such as survival analysis, Bayesian statistics, degradation modeling, time series analysis, and Monte Carlo simulation, to estimate the remaining useful life of assets.

The applications of residual life prediction span across various industries, including predictive maintenance, asset management, warranty management, risk assessment, and insurance, energy systems and infrastructure, and transportation and fleet management. Accurate estimation of residual life enables organizations to optimize maintenance strategies, improve asset performance, reduce downtime, and enhance operational efficiency.

The references provided throughout the chapter offer further reading and support for the concepts discussed. By understanding and applying the principles of residual life prediction, organizations can make informed decisions regarding asset maintenance, replacement, and resource allocation, leading to cost savings, improved reliability, and enhanced overall performance.

Residual life prediction continues to evolve with advancements in technology, such as sensor networks, IoT, and big data analytics. The future holds promising opportunities for further developments and enhancements in residual life prediction models and techniques.

Overall, this chapter aims to equip readers with a comprehensive understanding of residual life prediction, its methodologies, applications, and challenges. By leveraging the insights and recommendations presented, organizations can enhance their asset management practices, optimize maintenance strategies, and improve operational outcomes.

References

1. Ivanov, V., II, Dolgui, A., Sokolov, B.V., Asset lifecycle management: Recent advances and challenges. *Int. J. Prod. Res.*, 59, 1, 101–120, 2021.
2. Jardine, A.K., Lin, D., Banjevic, D., A review on machinery diagnostics and prognostics implementing condition-based maintenance. *Mech. Syst. Signal Process.*, 20, 7, 1483–1510, 2006.
3. Luo, Y., Guo, B., Li, M., Chen, C., Residual life prediction of key equipment, 2021.
4. Borges, A.M., Guedes, R., Pimentel, R., Barros, J.C., Remaining useful life prediction considering the occurrence of rare events: A review. *Reliab. Eng. Syst. Saf.*, 198, 106903, 2020.
5. Chen, S., Li, G., Li, Y., A comprehensive review of remaining useful life prediction methods for equipment maintenance decision support. *J. Manuf. Syst.*, 48, 144–152, 2018.
6. Li, R., Zhang, J., Wang, M., Chen, M., Remaining useful life prediction with data-driven deep hierarchical learning. *IEEE Trans. Ind. Electron.*, 66, 4, 3191–3201, 2019.
7. Xu, L., Chen, M., Hu, Q., Zhang, Z., Remaining useful life prediction of equipment with operating condition information. *Reliab. Eng. Syst. Saf.*, 215, 112955, 2021.
8. Zhang, X., Li, P., Zhang, X., Huang, S., Zhang, X., Remaining useful life prediction of complex equipment under small sample and imperfect repair. *Reliab. Eng. Syst. Saf.*, 198, 106913, 2020.

9. Ma, L., Yu, J., Zhou, D., A hybrid prognostic model for remaining useful life prediction based on deep learning and physics of failure. *Reliab. Eng. Syst. Saf.*, 193, 106598, 2020.
10. Nguyen, H.T., Liao, H., Ma, L., Prognostics and health management: A review on data-driven approaches. *Eng. Appl. Artif. Intell.*, 72, 294–315, 2018.
11. Wang, D., Li, Y., Zhang, C., Jiang, Y., Zhang, R., Deep learning models for remaining useful life prediction: A review and new perspectives. *Reliab. Eng. Syst. Saf.*, 214, 107612, 2021.
12. Zhang, X., Tian, Z., Liu, H., Li, P., An overview on ensemble methods for prognostics and health management. *Reliab. Eng. Syst. Saf.*, 188, 185–199, 2019.
13. Cai, X., Jardine, A.K., Lin, D., Remaining useful life prediction with degradation rate model using a degradation process governed by a Wiener process. *Reliab. Eng. Syst. Saf.*, 172, 49–57, 2018.
14. Chen, L., Huang, H.Z., Wang, W., Xu, Z., Remaining useful life prediction based on hierarchical Bayesian model with particle swarm optimization. *Reliab. Eng. Syst. Saf.*, 207, 107418, 2021.
15. Feng, Y., Liu, J., Li, W., Xie, M., Remaining useful life prediction of deteriorating systems using integrated health indices and degradation models. *Reliab. Eng. Syst. Saf.*, 197, 106930, 2020.
16. Huang, H., Ye, Q., Liu, S., Wang, S., Xie, M., Remaining useful life prediction based on a seasonal decomposition model with auto-regressive and integrated moving average errors. *Reliab. Eng. Syst. Saf.*, 199, 106837, 2020.
17. Afridi, M.J., Qureshi, T.A., Saeed, F., Remaining useful life prediction of mechanical systems based on LSTM neural networks. *Neural Comput. Appl.*, 33, 1, 165–174, 2021.
18. Javed, K., Muhammad, A., Zhang, Y., Khan, M.S., Mehmood, C.A., Remaining useful life prediction of rolling element bearings using deep random forest. *IEEE Access*, 8, 21923–21934, 2020.
19. Wang, W., Zhang, J., Tang, Y., Li, S., Zhang, C., Remaining useful life prediction for equipment subject to condition monitoring based on a hybrid deep learning method. *IEEE Trans. Ind. Inf.*, 14, 8, 3598–3607, 2016.
20. Wang, D., Li, Y., Zhang, C., Jiang, Y., Zhang, R., Deep learning models for remaining useful life prediction: A review and new perspectives. *Reliab. Eng. Syst. Saf.*, 214, 107612, 2018.
21. Bui, N., Nguyen, N., Chua, C., Xiao, G., A comprehensive review of remaining useful life prediction models: A data-driven approach. *J. Clean. Prod.*, 298, 126826, 2021.
22. Ding, R., Wang, Z., Lu, X., Chen, J., Residual useful life prediction for products under warranty using multilevel degradation models. *Reliab. Eng. Syst. Saf.*, 199, 106935, 2020.
23. Khan, U.R., Ullah, H., Khan, M.A., Ahmad, J., Mahmood, H., Naeem, M.A., Remaining useful life estimation in power plant steam turbine using deep learning techniques. *IEEE Access*, 7, 31100–31109, 2019.

24. Shi, Y., Wang, Z., Lu, C., Zhang, X., An integrated approach for asset management based on remaining useful life estimation and optimization. *J. Intell. Manuf.*, 30, 6, 2299–2312, 2019.
25. Song, J., Jiang, Y., Wang, H., Data-driven remaining useful life prediction for rolling element bearings under time-varying operating conditions. *Mech. Syst. Signal Process.*, 144, 106876, 2020.
26. Wang, W., Zhang, J., Tang, Y., Li, S., Zhang, C., Remaining useful life prediction for equipment subject to condition monitoring based on a hybrid deep learning method. *IEEE Trans. Ind. Inf.*, 14, 8, 3598–3607, 2018.
27. Wang, Y., Zhao, Y., Addepalli, S., Remaining useful life prediction using deep learning approaches: A review. *Procedia Manuf.*, 49, 81–88, 2020.
28. Rahmani AM, Azhir E, Ali S, Mohammadi M, Ahmed OH, Ghafour MY, Ahmed SH, Hosseinzadeh M. Artificial intelligence approaches and mechanisms for big data analytics: a systematic study. *PeerJ Comput. Sci.*, 14;7:e488, 2021.
29. Rashid, A.B. and Kausik, A.K., AI revolutionizing industries worldwide: A comprehensive overview of its diverse applications. *Hybrid Advances*, 100277, 2024.

6

Analysis of Vibration

Rajdeep Singh* **and Chandan Deep Singh**

Department of Mechanical Engineering, Punjabi University, Patiala, India

Abstract

Vibrations of machine tools limit quality and productivity and therefore, need to be analyzed. In principle, video-based vibration analysis using spatial phase information could reduce measuring effort significantly. However, the feasibility of these approaches for moving machine tool axes and relative movement between camera and object is yet to be shown. This paper investigates an approach for stabilizing linear axis movement to analyze vibration. It is shown that shaker-induced vibration on a linearly moving clamping tower can be derived for low frequencies. However, influences of actual cutting processes as well as compensation of unwanted camera vibration need further investigation.

Keywords: Vibration, frequency, bandwidth, linear axis, gear failure

Introduction

What is Vibration Analysis?

Vibration analysis is defined as a process for measuring the vibration levels and frequencies of machinery and then using that information to analyze how healthy the machines and their components are. While the inner-workings and formulas used to calculate various forms of vibration can get complicated, it all starts with using an accelerometer to measure vibration. Anytime a piece of machinery is running, it is making vibrations. An accelerometer attached to the machine generates a voltage signal that corresponds to the amount of vibration and the frequency of vibration

**Corresponding author*: rajdeepsingh.87@gmail.com

Chandan Deep Singh, Davinder Singh, Kanwal Jit Singh and Harleen Kaur (eds.) Condition-Based Maintenance and Residual Life Prediction, (101–114) © 2025 Scrivener Publishing LLC

the machine produces, usually the number of times per second or minute the vibration occurs (Demircioğlu and Çakır, 2023).

All data collected from the accelerometer go directly into a data collector (software) that records the signal as either amplitude vs. time (known as time waveform), amplitude vs. frequency (known as fast Fourier transform), or both. All of these data are analyzed by computer program algorithms, which in turn are analyzed by engineers or trained vibration analysts to determine the health of the machine and identify possible impending problems like looseness, unbalance, misalignment, and lubrication issues. Vibration analysis can detect problems (Shi, Yang and Wu, 2023) such as

- Imbalance,
- Bearing failures,
- Mechanical looseness,
- Misalignment,
- Resonance and natural frequencies,
- Electrical motor faults,
- Bent shafts,
- Gearbox failures,
- Empty space or bubbles (cavitation) in pumps, and
- Critical speeds.

The distributor and alignment and maintenance training provider VibrAlign uses the example of taking an industrial fan, removing a fan blade, and starting it up. As expected, the fan vibrates due to an unbalanced fan wheel. This unbalanced force will occur one time per revolution of the fan. Another example would be a damaged bearing track causing a bearing roller to generate vibration each time it contacts the spall (similar to a pothole on a highway). If three bearing rollers hit the spall per revolution, you should see a vibration signal of three times the fan's running speed (Sayyad, Ghugal and Kant, 2023).

Vibration Analysis Methodology

While accelerometers are still the most common tool used to collect vibration data, modern technology and improved sensor technology have allowed for non-contact, high-speed laser sensors that can detect issues accelerometers cannot. This allows for a more accurate and more localized analysis, and opens up vibration analysis to more methodologies. Vibration analysis is generally broken down into four principles, and each principle gives you specific information on the working conditions and features of the vibrating parts (Kecik, Smagala and Ciecieląg, 2023).

1. **Time domain:** When a vibration signal is picked up from a transducer (a device that converts a physical quantity into an electrical signal) and displayed on the screen of an oscilloscope, it is called a waveform. This signal is in the time domain. The time domain is amplitude plotted against time. While most machine vibration issues are detected using spectrum analysis, some types are more easily seen in waveform.
2. **Frequency domain:** When the waveform discussed earlier is subjected to spectrum analysis, the end result is a picture of frequency vs. amplitude, known as a spectrum. The spectrum is in the frequency domain like the vibration is in the time domain. Most in-depth analysis of machinery vibration are done in the frequency domain or using spectrum analysis.
3. **Joint domain:** Because vibration signals vary with time, calculating more than one spectrum at once can be useful. To do this, a joint time technique called Gabor–Wigner wavelet can be utilized. This technique calculates variations of the fast Fourier transform (discussed below), including short-time Fourier transform (STFT).
4. **Modal analysis:** Modal analysis takes measured frequency response functions of a piece of machinery and puts them into a computer model. The computer model can be displayed with animations of all the different vibration modes. The model can be adjusted by either adding to or taking away things like mass or stiffness to see the effects.

Outside these four basic principles lie numerous forms of analysis, calculations, and algorithms used to determine different aspects of vibration analysis. These (Vu *et al.*, 2023) include the following:

Centrifugal pump is widely used in the water system of nuclear power plant. It is a key equipment of pressurized water reactor in a nuclear power plant. Its operation reliability and stability are quite important to the safety and economy of power plants. Due to the diversity and complexity of nuclear power plant systems and working conditions, centrifugal pumps usually operate and work on the condition that deviate from the design conditions over a long period of time. When the centrifugal pump operates

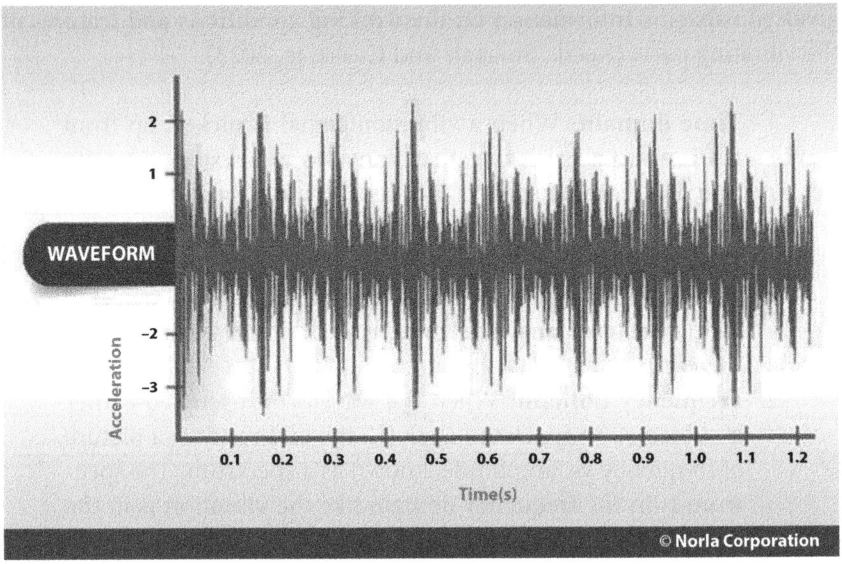

Figure 6.1 FFT waveform.

at low flow conditions, complex unstable flow phenomena such as fluid separation, reflux, rotating stall, cavitation, and pressure pulsation will occur in the pumps (Frankowski, Chady and Zieliński, 2021).

- **Time waveform:** A time waveform is acceleration vs. time displayed as tables and plots. Time waveforms show a short time sample of raw vibration, revealing clues to the condition of machinery not always clear in the frequency spectrum. A method of employing time waveform vibration signals as a vibration analysis tool is by using FFT.
- **Fast Fourier transform (FFT):** FFT is defined as an algorithm used to calculate a spectrum from a time waveform. In other words, it is a calculation intended to break down a signal into all its frequencies. If you will recall time domain and frequency domain discussed above (see Figure 6.1), FFT converts a signal from the time domain into the frequency domain. Fast Fourier transform is most often used for detecting machine faults like misalignment or unbalance.
- **Phase measurement:** When talking about vibration analysis, phase is a relative time difference between two signals measured in units of angle as opposed to time. It only works

if the two signals being compared are of the same frequency. Phase measurement is used in tandem with FFT to decipher machine faults like loose parts, misalignment, and unbalance.

- **Order analysis:** This is a variation of FFT analysis and is mostly used to quantify vibrations of machines with varying revolutions per minute (RPM). In other words, order analysis is frequency analysis where the spectrum's frequency axis is shown in orders of RPM rather than hertz. The term "orders" refers to a frequency that is a multiple of a reference rotational speed. For example, if a vibration signal is equal to twice the frequency of the motor's rotation, the order is two (Selim *et al.*, 2023).
- **Power spectral density (PSD):** This is calculated by multiplying the amplitude from the FFT by its different forms to normalize it with the frequency bin width (bin width refers to the grouped x-axis values) (see Figure 6.2). Think of PSD as looking at "random" vibrations or motion at many different frequencies. PSD accurately compares random vibration signals that have different signal lengths (REN *et al.*, 2023).

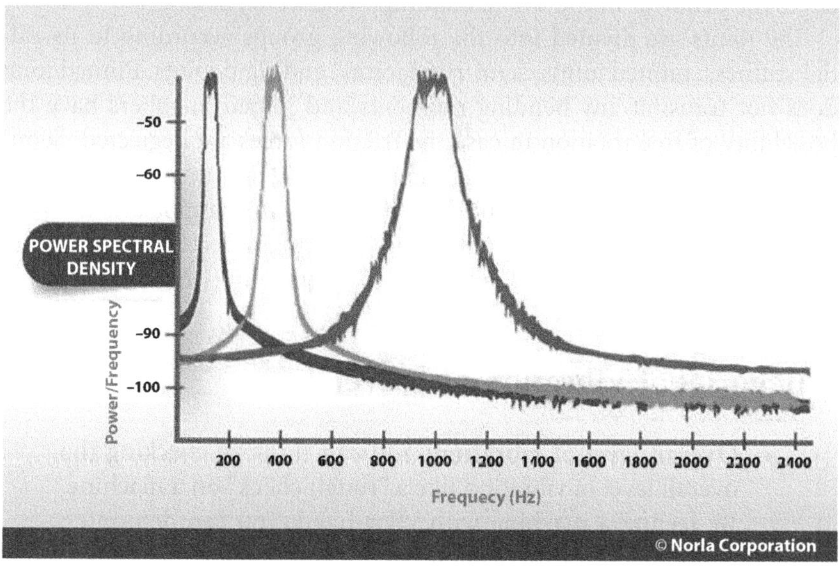

Figure 6.2 Power spectral density.

- **Envelope analysis:** Envelope analysis is a form of vibration analysis that can detect impacts with very low energy often hidden by other vibration signals. It is a popular diagnostic tool for damaged gear teeth and roller bearings (Saito and Suzuki, 2022).
- **Orbit:** The orbit is defined as a plot of a sleeve journal bearing centerline. It is measured by placing two probes in the bearing housing 90 degrees apart. Data from these probes can be displayed digitally and used to detect shaft vibrations caused by oil whirl—oil whirling around inside, causing the journal to move.
- **Resonance analysis:** This identifies all the natural vibrations and frequencies in machines. The presence of resonance means high vibration, which could reach damaging levels (Tan and Tang, 2023).

The structural health of a building is determined by the structural health of its separate members and joints. The behavior of the structural members and joints should be under control during the whole service life of the building. Distribution of internal forces and deformations in structural members and joints depend on the joint's stiffness. Stiffness is a major parameter, which determines the structural health of the joints and influences the behavior of the whole building (Vaka *et al.*, 2022).

The joints are divided into the following groups according to its initial stiffness: pinned joints, semi-rigid joints, and rigid joints. Pinned joint does not transmit any bending moments and joined members have the possibility of free rotation in case the friction forces are neglected. Semi-rigid joint transfers some bending moments by its sub-members and connections and its behavior should be considered in the global analysis of whole building structures. Members are rigidly connected to each other and fully transmit bending moments in the rigid joint (REN *et al.*, 2023).

Categories of Vibration Measurement

- **Overall level of vibration:** You can think of checking the overall level of vibration like a "rough check" on a machine. By feeling a machine with your hand, you can determine a general sense of whether it is running roughly over a wide frequency band. This initial check is best on rotating machinery, in particular, high-speed machines. It is not usually applicable to reciprocating machines.

- **Spectral analysis of vibration:** Spectral analysis is the process of transforming a signal from the time domain to the frequency domain. It is often done using FFT. The signal is analyzed to determine any substantial frequencies coming from the machine's components. Where there is a peak in frequency signal that is the likely source of vibration. Common applications for spectral analysis include the rotational speed of a shaft or how often tooth meshing occurs on a pair of gear wheels (Klee *et al.*, 2020).
- **Discrete frequency monitoring:** If you need to monitor a specific component within a machine, discrete frequency monitoring measures the vibration level being generated at a particular frequency, which that component would be expected to generate. For example, if you want to look into a certain shaft in a machine, you would turn the monitoring to that machine's rotational speed. Discrete frequency is calculated using the FFT algorithm.
- **Shock pulse monitoring:** Shock pulse monitoring is a predictive maintenance technique that monitors rolling-element bearings with a hand-held instrument. The hand-held instrument gives off a natural frequency that is excited by shocks or vibrations generated by rolling bearings. In other words, when two pieces of metal touch each other while in motion, shock waves develop from the impact, which travels through the metal. This shockwave is used in shock pulse monitoring (Frankowski, Chady and Zieliński, 2021).
- **Kurtosis measurement:** Kurtosis gives you a measure of the "spikiness" of a random signal. Signals with a higher kurtosis value have more peaks that are greater than three times the signal's root mean square (RMS) value. In vibration analysis, kurtosis is used to monitor fatigue development in rolling bearings with a simple instrument.
- **Signal averaging:** Since signals change with time, signal averaging is important in spectrum analysis because it determines the level of the signal at each frequency. It is particularly important for low-frequency measurements because they need a longer averaging time to get a statically accurate

estimate of the spectrum. Signal averaging is often used in the monitoring of a gear in relation to its rotational speed. In this example, signal averaging will show you the cyclic action of each tooth in the gear. If a tooth has a large crack, it would be detected due to its increased flexibility (Rallu *et al.*, 2023).

- **Cepstrum analysis:** Originally invented to characterize seismic echoes produced by earthquakes and bomb explosions, cepstrum is used to look at the repeated patterns in a spectrum. Repeated patterns in the spectrum are sensed as one or two components in the cepstrum with multiple sets of sidebands, which can be confusing. The cepstrum separates those sidebands like the way the spectrum separates repetitive time patterns in the waveform. Cepstrum analysis is often utilized to look at interactions between the rotational frequency of bladed rotors and the blade passing frequency. Another example is examining gear tooth meshing frequencies and gear rotational speeds (Pham, Tran and Nguyen, 2023).

Vibration Analysis: Measurement Parameters

All these vibration analysis techniques help to identify three major parameters: acceleration, velocity (RMS), and displacement. Each of these parameters emphasizes certain frequency ranges in their own way and can be analyzed together to diagnose issues. Let's take a look at each parameter (Moreira *et al.*, 2023).

Compared to metal, the application of composites has grown in the previous decades due to required light weight and high strength structures and design ability to construct the structures with high-stress tolerance in an arbitrary direction in modern engineering. More information about laminated composite structures and functionally graded materials (FGMs) can be observed with a deep consideration to the complementary references (Zhao *et al.*, 2022).

- **Acceleration:** Acceleration places greater importance on high frequencies. An acceleration signal is not exclusive, however. The acceleration signal can be converted to velocity or displacement.
- **Displacement:** Just like acceleration places greater importance on high frequencies, displacement looks at

low frequencies. Displacement measurements are generally only used when examining the broad picture of mechanical vibrations. You might use displacement to discover unbalance in a rotating part due to a significant amount of displacement at the rotational frequencies of the machine's shaft.

- **Velocity:** Velocity is related to the destructive force of vibration, making it the most important parameter. It places equal importance on both high and low frequencies. Usually, the RMS value of velocity (measured in the range of 10 to 10,000 Hz) shows the best sign of vibration severity. RMS is calculated by multiplying peak amplitude by 0.707 (Selim and Althobaiti, 2022).

Below is an example of what acceleration, displacement, and velocity look like on the same signal (see Figure 6.3). You can see some peaks at the same frequencies, but each has different amplitudes. This is a good visual of how each parameter assigns different importance to frequency ranges (Lee, Choi and Lee, 2023).

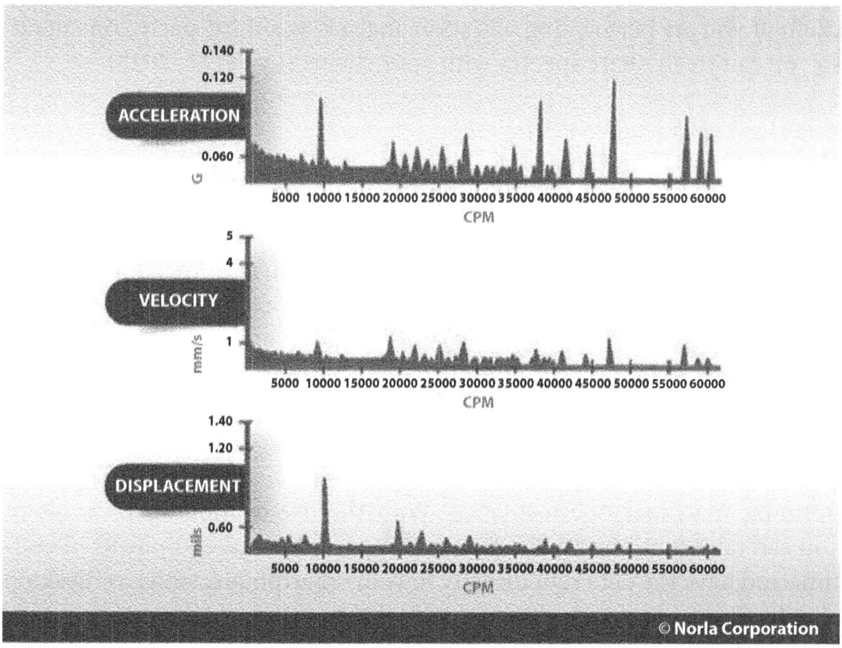

Figure 6.3 Acceleration, velocity, and displacement.

Vibration Analysis: Tools and Technology

Advanced technology, particularly advances in wireless technology, has greatly improved how vibration analysts collect, interpret, and share data. Today, vibration analyzers are extremely portable, communicate with smartphones and tablets in real-time, and can generate FFT in extremely high resolution. Many vibration instrument companies develop their own apps to communicate with each other.

Another form of advanced technology you will see with vibration analysis interpretation instruments is operating deflection shapes (ODS) of 3D simulations of machinery vibrations. In a nutshell, this type of software exaggerates vibration-induced movements in a 3D model so you can visualize the forces impacting your machine while it is running.

Some companies that offer vibration analysis instruments also offer databases with thousands of bearing fault frequencies preloaded to help you identify certain fault frequencies for your bearings. Some software can continuously monitor the geometry of your rolling elements and warn you when possible premature failures may occur.

As with most advanced technology, the majority of vibration analysis data are automatically uploaded to the cloud and are available on your mobile device, computer or directly from your browser. This is especially useful if you are performing vibration analysis as a third-party consultant, so you can freely share spectra with your clients (Lyu *et al.*, 2023).

Benefits of Continuous Vibration Monitoring

The methods and tools discussed in this chapter are not only great for determining what is wrong with a piece of equipment or machinery (reactive), but they can also be used to catch issues before they cause significant downtime (proactive). Using vibration analysis and monitoring enables you to look quantitatively into structural weakness or looseness, rotating component looseness, and whether resonance is present (Robin *et al.*, 2022).

If implemented properly, continuous vibration monitoring helps you optimize machinery performance. With the use of modern technology, you can take continuous vibration readings on various equipment in real-time and have the data sent directly to your smartphone, tablet, or desktop via the cloud.

- **Monitor critical equipment:** Critical equipment is any piece of equipment or machine that could cause you to take a big financial hit if a failure were to occur. Continuous vibration monitoring helps detect discrepancies in the vibration spectrum, which can reveal lubrication issues and bearing defects well before major issues appear.
- **Monitor heavily used equipment:** Many plants operate 24/7, only stopping monthly or quarterly for routine maintenance. Stopping more than this can cost the plant a significant amount of money. Online continuous vibration monitoring helps monitor the condition of heavily used machinery or troubled machinery and sends alerts when that condition changes.
- **Monitor difficult-to-access equipment:** Performing maintenance on equipment located in hard-to-reach places is difficult. Machines on rooftops, cooling towers, and those operating in high-temperature areas can be continuously monitored for vibration abnormalities, allowing maintenance to be done at a convenient time. This prevents unplanned downtime and keeps maintenance staff from accessing these locations unnecessarily (Kuriakose and Sreehari, 2023).

References

Demircioğlu, U. and Çakır, M.T., Free vibration analysis of curved asymmetric sandwich structures. *J. Eng. Res.*, 11, 2, 100036, 2023, doi: 10.1016/j.jer.2023.100036.

Frankowski, P.K., Chady, T., Zieliński, A., Magnetic force induced vibration evaluation (M5) method for frequency analysis of rebar-debonding in reinforced concrete. *Meas.: J. Int. Meas. Confed.*, 182, 124–138, February, 2021, doi: 10.1016/j.measurement.2021.109655.

Kecik, K., Smagala, A., Ciecieląg, K., Diagnosis of angular contact ball bearing defects based on recurrence diagrams and quantification analysis of vibration signals. *Meas.: J. Int. Meas. Confed.*, 216, April, 112963, 2023. doi: 10.1016/j.measurement.2023.112963.

Klee, B. *et al.*, Vibration analysis of moving machine tool axes based on phase information in video data. *Procedia CIRP*, 88, March, 364–368, 2020, doi: 10.1016/j.procir.2020.05.063.

Kuriakose, V.M. and Sreehari, V.M., Composites Part C : Open Access Vibration and flutter analysis of damaged composite plates under thermal environment and its passive control using piezoelectric patches. *Composites, Part C: Open Access*, 11, May, 100361, 2023, doi: 10.1016/j.jcomc.2023.100361.

Lee, S., Choi, W., Lee, D.H., 'The vibration knows who you are! A further analysis on usable authentication for smartwatch users. *Comput. Secur.*, 125, 103040, 2023, doi: 10.1016/j.cose.2022.103040.

Lyu, F. *et al.*, Vibration characteristics analysis of high-pressure long-distance dense paste pipeline transporting with hydraulic active check valve. *Alexandria Eng. J.*, 71, 669–677, 2023, doi: 10.1016/j.aej.2023.03.086.

Moreira, J.A. *et al.*, Assessment of layerwise user-elements in Abaqus for static and free vibration analysis of variable stiffness composite laminates. *Compos. Struct.*, 303, April 2022, 2023, doi: 10.1016/j.compstruct.2022.116291.

Pham, Q.H., Tran, T.T., Nguyen, P.C., Nonlocal free vibration of functionally graded porous nanoplates using higher-order isogeometric analysis and ANN prediction. *Alexandria Eng. J.*, 66, 651–667, 2023, doi: 10.1016/j.aej.2022.12.023.

Rallu, A. *et al.*, Vibrations induced by tunnel boring machine in urban areas: *In situ* measurements and methodology of analysis. *J. Rock Mech. Geotech. Eng.*, 15, 1, 130–145, 2023, doi: 10.1016/j.jrmge.2022.02.014.

REN, S. *et al.*, Multi-probe linear fitting and time of arrival linear correction method to analyze blade vibration based on blade tip timing without once-per-revolution. *Chin. J. Aeronaut.*, 36, 1, 290–310, 2023, doi: 10.1016/j.cja.2022.06.006.

Robin, G. *et al.*, Non-linear vibration analysis of visco-elastically damped composite structures by multilevel finite elements and asymptotic numerical method. *Composites, Part C: Open Access*, 7, 100240, 2022, doi: 10.1016/j.jcomc.2022.100240.

Suzuki, T., Forced response vibration analysis of induction motor stators induced by electromagnetic forces. *IFAC-PapersOnLine*, 55, 27, 155–159, 2022, doi: 10.1016/j.ifacol.2022.10.504.

Sayyad, A.S., Ghugal, Y.M., Kant, T., Higher-order static and free vibration analysis of doubly-curved FGM sandwich shallow shells. *Forces Mech.*, 11, April, 100194, 2023, doi: 10.1016/j.finmec.2023.100194.

Selim, M.M. *et al.*, Torsional vibrational analysis of irregular single-walled carbon nanotube with elastic-support boundary conditions. *J. Mater. Res. Technol.*, 24, 215–222, 2023, doi: 10.1016/j.jmrt.2023.02.230.

Selim, M.M. and Althobaiti, S., Wave-based method for longitudinal vibrational analysis of irregular single-walled carbon nanotube with elastic-support boundary conditions. *Alexandria Eng. J.*, 61, 12, 12129–12138, 2022, doi: 10.1016/j.aej.2022.06.001.

Shi, Y.H., Yang, D.Q., Wu, W.W., Numerical analysis method of ship-ice collision-induced vibration of the polar transport vessel based on the full coupling of ship-ice-water-air. *J. Ocean. Eng. Sci.*, xxxx, 2023, doi: 10.1016/j.joes.2023.04.002.

Tan, X., and Tang, Y.Q., Free vibration analysis of Timoshenko pipes with fixed boundary conditions conveying high velocity fluid. *Heliyon*, 9, 4, 2003, e14716, doi: 10.1016/j.heliyon.2023.e14716.

Vaka, V. *et al.*, Steady-state and transient vibration analysis of a thermally loaded power law-based functionally graded rotor system with induced porosities. *Forces Mech.*, 9, July, 100123, 2022, doi: 10.1016/j.finmec.2022.100123.

Vu, N.A. *et al.*, Third-order isogeometric analysis for vibration characteristics of FGP plates in the thermal environment supported by Kerr foundation. *Case Stud. Therm. Eng.*, 45, January, 102890, 2023, doi: 10.1016/j.csite.2023.102890.

Zhao, G. *et al.*, Vibration analysis of size dependent micro FML cylindrical shell reinforced by CNTs based on modified couple stress theory. *Arab. J. Chem.*, 15, 10, 104115, 2022, doi: 10.1016/j.arabjc.2022.104115.

7

Modeling for Vibration

Rajdeep Singh* **and Chandan Deep Singh**

Department of Mechanical Engineering, Punjabi University, Patiala, India

Abstract

The application of modeling approaches for vibration analysis is the main topic of this chapter. To create and analyze models and anticipate various forms of engine generated vibrations, modeling-based analysis is heavily utilized. In terms of Eigenvalues, it is also helpful for forecasting nodal displacements, stress-induced frequencies, and mode shapes. The examination of engine-developed vibrations first focuses on finite element analysis as the most widely utilized investigative technique. ANSYS, HyperMesh ABAQUS solver, COMSOL, and HyperMesh OptiStruct solver have been used in recent finite element analysis (FEA) simulation research. In addition to FEA, researchers also perform mathematical modeling of generated models using software such as MATLAB, Directed Tabu Search (DTS), leap frog approach, etc. Academicians, scientists, and researchers looking to choose the most appropriate and practical technique for engine vibration analysis will find this comprehensive evaluation of the numerous modeling techniques for engine vibration analysis to be helpful.

Keywords: Vibration analysis, simulation, FEA, mathematical modeling

7.1 Introduction

Lowering the fuel consumption of automobiles is one of the main problems facing modern automotive development. Taking into account the desirable characteristics of a sports car, which essentially include its performance, speed, and capacity for racing. Reducing a vehicle's bulk is a simple approach to improve its fuel economy. Reduced masses, however, cause more issues with vibrations and overall dynamical behavior.

**Corresponding author*: rajdeepsingh.87@gmail.com

Chandan Deep Singh, Davinder Singh, Kanwal Jit Singh and Harleen Kaur (eds.) Condition-Based Maintenance and Residual Life Prediction, (115–130)

More vibrations result in more wear and a decrease in comfort, and since comfort is a crucial quality factor that might affect a customer's purchasing decision, this is a problem. The engine causes the majority of the vibrations in a vehicle. Researchers and engine makers have identified some parts that cause vibration. Engine components such as the engine block, engine head, piston, connecting rod, crankshaft, flywheel, and camshaft, as well as valves, manifolds, pulleys, etc., are the primary vibration-initiating sources. Unbalance in the engines leads to increased strains in the rotating system due to vibration because of flawed design and subpar workmanship. Additionally, the vibration hastens the wear of gears and bearings, which are contact parts. Unwanted vibrations may cause parts to become loose and frame and engine mounting to become damaged as a result of stress, which could lead to early failure and a shorter product lifespan (Galván *et al.*, 2022).

In essence, simulation is a procedure that analyses any invention or current design before it is used in practice. Static, modal, and dynamic analyses are used in simulation to determine design and application parameters including stress, displacement, fatigue life, thermal conductivity, and so on. For many fields, including mechanical engineering, electrical engineering, civil engineering, medical science, etc., there are currently numerous simulation and computer aided design (CAD) software options accessible. In an internal combustion (IC) engines, simulation is typically used to forecast combustion chamber behavior as well as fluid flow (gas, mixture, temperature, and vibration) (Serdjuks *et al.*, 2021).

7.2 Modeling Techniques for Vibration Analysis

Numerous researchers are employing various types of approaches to design, analyze, and optimize engine installation. To provide accurate research results, researchers have used a variety of simulation software programs or mathematical modeling techniques. This review is offered here through an examination of key works that are available in the form of articles published in journals, conferences, reports from numerous working groups, and many websites. The focus of the current review article is mostly on important studies conducted in previous years. To gain an understanding of the methodology used by researchers, the types of analyses they conducted, and their important findings, a critical synopsis of research on engine mount analysis is offered below. The review is divided into categories based on the modeling approach that the researcher used to analyze the engine mount bracket and engine vibration, and the classification is carried out as follows (Khazaee *et al.*, 2022).

1) Swanson Analysis Systems, Inc (ANSYS) Simulation
2) HyperMesh ABAQUS Simulation
3) HyperMesh OptiStruct Simulation
4) COMSOL Multiphysics Simulation
5) Mathematical Modeling
   - Matrix laboratory (MATLAB) Simulation
   - Miscellaneous Techniques

7.2.1 ANSYS Simulation

A finite element analysis tool for structural analysis, including linear, nonlinear, and dynamic investigations, is ANSYS Mechanical simulation. For a variety of mechanical design issues, this computer simulation tool supports material models, equation solvers, and finite elements to represent behavior. Along with acoustics, piezoelectric, thermal-structural, and thermo-electric analysis, ANSYS Mechanical also offers thermal analysis. For modeling the behavior of materials to short-duration severe loadings from impact, high pressure, or explosions, ANSYS is a computer simulation tool.

Abbasi *et al.* (2022) used FEA to perform a static structural and modal study on the engine mounting bracket. Using FEA software ANSYS, materials such as aluminum (Al), magnesium (Mg), and cast iron (CI) were examined through the natural frequency of the engine mounting bracket. The analysis's findings reveal a consistent relationship between natural frequency versus stress and stress versus deformations for a variety of materials with varied mesh sizes. The analysis's findings showed that the Mg alloy bracket has the highest natural frequency, followed by the Al alloy and the CI.

The researcher performed modal analyses on the engine bracket using the same FEA program to determine the natural frequencies of the different materials under consideration—Al alloy, Mg alloy, and grey CI. Through the use of ANSYS, they examined the material stress and deformation. According to the analysis's findings, the natural frequency of magnesium alloy is superior to that of aluminum alloy and grey cast iron engine brackets.

Saidin *et al.* (2022) did a study on the engine power-train mounting system utilizing FEA software using ANSYS, which is comparable to a composite body of one elastic body and one inertial body. In the equivalent analysis model method, the mounting components are reduced to a single spring along its three main elastic axes, and the mass and rotational inertia of the drivetrain are precisely embodied. The analysis's findings for energy decoupling intensities and vibration modes in the designated directions.

Adopting the same methodology of FEA static simulation and experimental analysis was done by Abouelregal and Alesemi (2022) on conventional materials such as gray cast iron, Al alloy, and Mg alloy. In this case, ANSYS was also used for static and modal analysis.

The simulation-based analysis was followed by an experimental examination using the same materials. According to the analysis's findings, aluminum alloy is the best material because its obtained frequencies are higher than those of the other two materials and its stress is within its yield value. Later, the mounting bracket's weight was optimized by being reduced by 21.8%. Later, a significant static and modal investigation using a square cross-section engine mounting bracket was carried out by Medolago and Melzi (2021).

Engine mounting brackets were analyzed using FEA software ANSYS by modifying the component's material (Al alloy and Mg alloy) (X. Wang *et al.*, 2021). The results of their study indicated that magnesium alloy is the optimum material for engines with square cross sections.

Chai *et al.* (2020) carried out fundamental research on the mathematical modeling and shape optimization of the Ashok Leyland 1612 truck's front damper mount. With limits on stiffness, strength, and natural frequency, the researchers worked to optimize the damper bracket. The geometric model was created using the modeling program SOLIDWORKS, and additional analysis was done using ANSYS. The outcome demonstrated that the new, optimized design had less displacement and stress. The bracket's optimized mass is lowered by 21.15%.

Similarly, Chen *et al.* (2020) conducted static and modal analysis of the engine mounting bracket. They examined the natural frequency of the engine bracket analytically and by developing the model and self-excitation frequency of the engine bracket. ERW-1, ERW-2, aluminum, and magnesium alloys for the engine mount bracket. For analysis, ANSYS was used as FEA solver. Shape optimization and material optimization were done on different materials. The results showed that the best material for the engine bracket was ERW-2, which is higher in excitation frequency.

Demirel and Kayran (2019) conducted static as well as modal analysis of an engine mounting bracket. The FEA analysis was done using ANSYS 15.0. The result obtained through the analysis showed a reduction in the weight of the proposed model of the engine mounting bracket by 12.5% while maintaining an acceptable level of yield stress and harmonic response.

Kim *et al.* (2018) optimized the Mahindra vehicle's engine mounting bracket. Without altering the original engine's supporting locations or types of support, the researcher created and analyzed an engine mounting bracket. In all circumstances, geometric modeling was carried out using CATIA V5 and was finished using the FEA program ANSYS. According to the study's findings, the maximum stress is 65.16 N/mm^2, which is within the safe limit, and the overall weight is decreased by 15% without compromising the structure's strength.

7.2.2 ABAQUS Simulation

For the modeling and analysis of mechanical parts and assemblies, as well as for visualizing the outcomes of finite element analysis, the HyperMesh solver software ABAQUS is utilized. In the ABAQUS/Viewer product, a portion of ABAQUS, only the post-processing module can be started individually. The automobile, aerospace, and industrial products industries all use ABAQUS.

Yalagi *et al.* (2019) performed modal and static analysis of an engine mount bracket using FEA tools. Using ANSYS, they carried out analyses on three different materials. After material assignment, HyperMesh was used for discretization, followed by the solver ABAQUS for analysis. For the engine mount bracket, cast iron, wrought iron, and mild steel were employed. The findings indicate that mild steel performs better than the other two examined materials.

7.2.3 HyperMesh OptiStruct Solver Simulation

The HyperMesh contemporary solver known as OptiStruct is employed for structural analysis, both linear and nonlinear, under static and dynamic

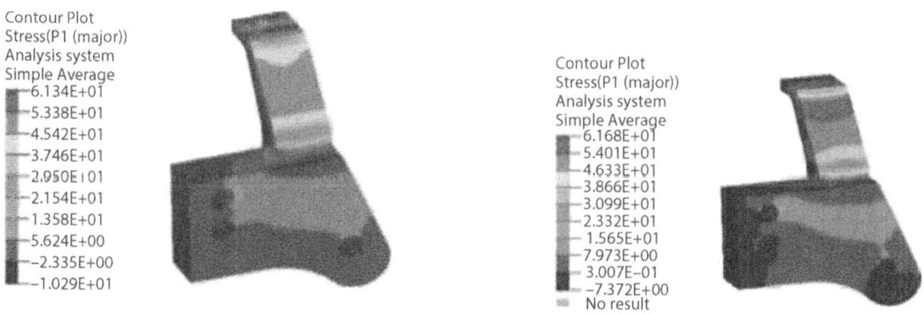

Figure 7.1 Stress optimization before and after using HyperMesh, Serrano *et al.* (2019).

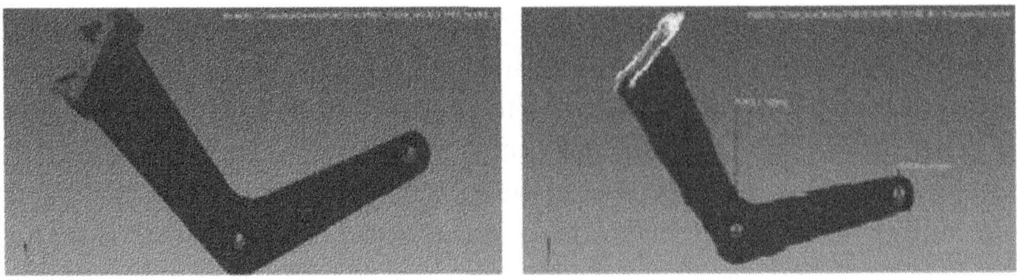

Figure 7.2 The meshed model and the boundary condition model.

loadings. It offers a structural design and optimization solution. This solution is used all over the world to evaluate and optimize structures for their NVH (noise, vibration, and harshness) characteristics as well as their strength and durability.

Static and vibration analysis of the engine mounting bracket for TMX 20-2 was done by Serrano *et al.* (2019) using the OptiStruct solver. The engine mounting was modeled in Unigraphics (NX7.5) and imported into HyperMesh. Strain gauges were placed where the brackets were failing in the field. The result obtained from the stress difference between physical testing and the simulation approach shows a deviation of about 18% and failure occurs where the ribs end on the bracket (see Figure 7.1).

Another group of researcher (Bertini *et al.*, 2018) conducted topology optimization in an engine mounting bracket through Finite Element Analysis using OptiStruct 10.0 solver. The results obtained after topology optimizations showed a stress of 61.34 N/mm^2 and displacement up to 0.357mm, satisfying the designing criteria. Furthermore, the researchers also reduced the mass by 0.22 kg per component.

Modal analysis and mass optimization were done by Ruocco *et al.* (2018) on an engine mount bracket for an formula SAE (FSAE) car. CAD modeling was done using the FEA software Creo 2.0 and discretization was done in HyperMesh 11, followed by analysis with the FEA solver OptiStruct, After designing, optimization was done on the engine mount bracket. The obtained results for the engine mounting bracket showed a 30.5% maximum deflection in the worst loading case. The maximum von Mises stress increased by 29.65%, maintaining a factor of safety of 3.3. The researchers optimized the weight by 11.64%. Figures 7.2a and b show the meshed model and the boundary condition model of the engine mounting bracket assembly.

7.2.4 COMSOL Simulation

COMSOL Multiphysics includes multiphysics simulation and finite element analysis tools. It employs coupled partial differential equation (PDE) systems and standard physics-based user interfaces. The structural mechanics module, which includes mechanical structures subject to static or dynamic loads, can benefit from the study performed by this software. Numerous other analysis types, including stationary, transient, eigenmode/modal, parametric, quasi-static, frequency-response, buckling, and pre-stressed, can be performed using it.

To reduce weight, the researchers (Okeke *et al.*, 2018) conducted FEA analysis on different materials for the engine mount bracket. The geometric model of the engine mount bracket was done on the CATIA V16 tool. Using the COMSOL Multiphysics 4.2 version software, the researchers applied thermo-mechanical vibration analysis to computationally test the component in its isotropic state. AlSiC Composite Material and Traditional Grey Cast Iron (GCI) were two materials that were examined for study using the same model. The outcome demonstrates that the engine mounting bracket's almunium silicon carbide (AlSiC) Composite material offers improved stability, consistent material behavior, and better performance. Additionally, the component's weight is decreased by 60%. During the thermo-mechanical vibration study using COMSOL Multiphysics software, Figure 7.3 illustrates the relationship between stress and frequency, and Figure 7.4 illustrates the link between displacement and frequency.

Y. Q. Wang *et al.* (2018) used COMSOL Multiphysics to do a structural analysis of the power-train mounting bracket. Starting with a CAD model

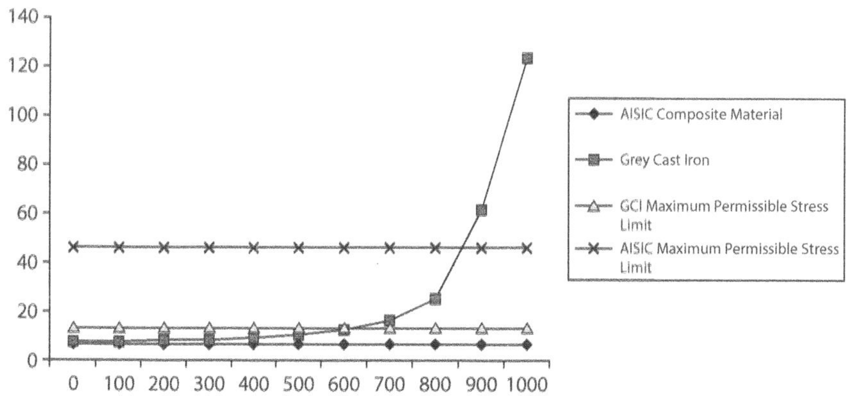

Figure 7.3 Stress versus frequency graph.

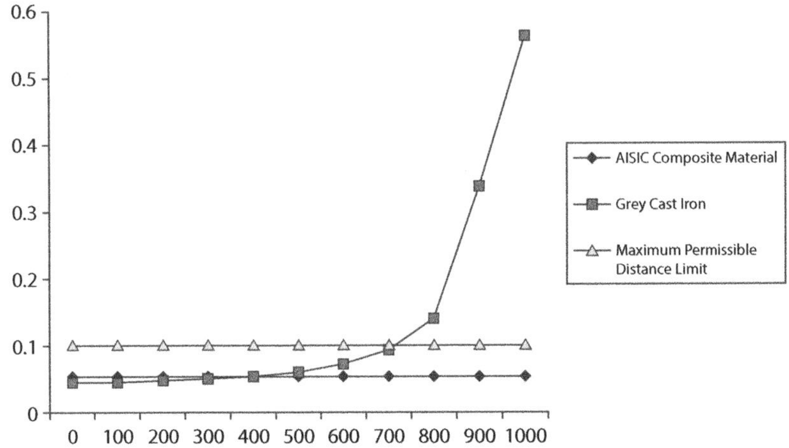

Figure 7.4 Displacement versus frequency graph.

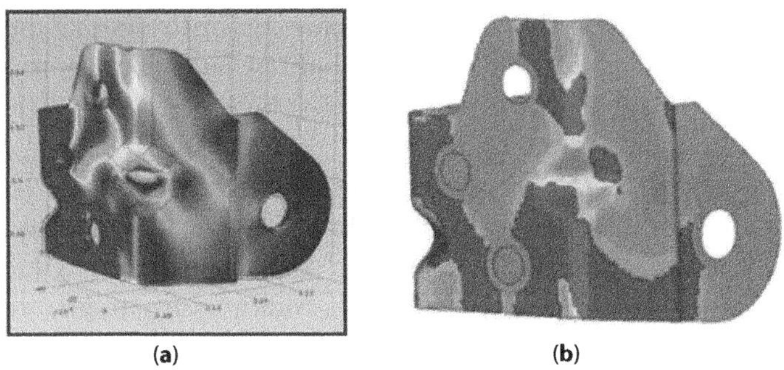

Figure 7.5 Stress plot: (a) COMSOL multiphysics model (b) HyperMesh model.

imported from CATIA V5R21, the COMSOL analysis is performed. The meshing element chosen is the tetrahedron. The content is chosen from the collection of "built-in" materials in COMSOL. The bracket is tested under a variety of load circumstances, such as 3-2-1 g weights, 3 g downward unbalanced torque, etc. It is optimized to ensure that the stresses it generates and the deflection of the bracket fall within acceptable bounds. Preloads on bolts are also taken into account. The results of the HyperMesh study were used to validate the COMSOL results. The correlation was extremely good in terms of values of stresses, stress patterns, and displacement. A comparison of stress plots for both analyses is shown in Figures 7.5a and b.

7.2.5 Mathematical Modeling

A mathematical model is a description of any system characterized by mathematical concepts and terminology. Mathematical modeling is the process of creating a mathematical model. Engineering and the natural sciences both use mathematical models. A model can aid in providing a system's specifics, studying the effects of various components, and making behavior predictions.

7.2.5.1 MATLAB Simulation

MATLAB is a multi-paradigm numerical computing environment and fourth-generation programming language that consists of two word matrices and a laboratory. Matrix manipulation, function and data visualization, algorithm implementation, user interface construction, and communication with programs written in other languages, such as C, C++, C#, Java, Fortran, and Python, are all possible in the proprietary programming language known as MATLAB.

Jabaraj and Jaafar (2013) used MATLAB to construct an optimization program for the design of a particular vehicle mounting system. Powertrain mounting system's dynamic model was created and optimized for energy decoupling. An FEA tool called automated dynamic analysis of mechanical systems (MSC.ADAMS/VIEW) was used to do a modal simulation analysis. Following optimal design, the optimized system's first-order mode and energy distribution of the frequency bands were within a suitable range.

The obtained results show a maximum frequency of 17Hz and the lowest frequency of 7Hz. Therefore, a high degree of decoupling of the system was achieved. Static and modal analysis of the engine mounting bracket was done by Nazemnezhad *et al.* (2017). They used a square cross-section for aluminum alloy and magnesium alloy. The analysis was done using an FEA tool followed by a comparison of both materials for the engine mounting bracket. They concluded that magnesium alloy is better than aluminum alloy for various operations.

The researchers (Gao *et al.*, 2015) studied the linear characterization of the engine mount and body mount for crash analysis with the help of MATLAB simulations. The mounts are treated at the component level, and mathematical models are evaluated to get the required characteristics. The mounts are modeled as spring and damper system subjected to impact loading that occurs during crash events. The approximations of input pulse were described mathematically, which then serves to find the characteristics of the mounts. The change in the characteristics of mounts with the change in the velocity of impact was also studied.

Through analytical computations for no excitation, half-sine excitation, and Fourier equivalent wave, the following results were obtained:

1) The displacement predicted by half-sine excitation is equivalent to the Fourier equivalent wave.
2) The breakage of the mount may occur after it has displaced as predicted by half-sine excitation.
3) The boundary conditions are well satisfied by the model, and thus the mathematical model can be used in the future as s standard tool to predict the deformations and velocity of the mounts at the initial stage.
4) The maximum displacement always occurs when the velocity is zero, which is an absolute theoretical condition.

7.2.5.2 Miscellaneous Techniques

Some researchers used distinct and customized type mathematical models for the analysis and optimization of engine mountings.

Tang *et al.* (2018) determined the most appropriate model for the optimization of Hydraulic Engine Mounts (HEMs). Hydraulic engine mounts are optimized to improve vehicle ride comfort using a global optimization method called the Directed Tabu Search (DTS) method. A full-vehicle model was used to evaluate the optimization results and to determine the optimized HEMs in the simplified models. Through the analysis, they

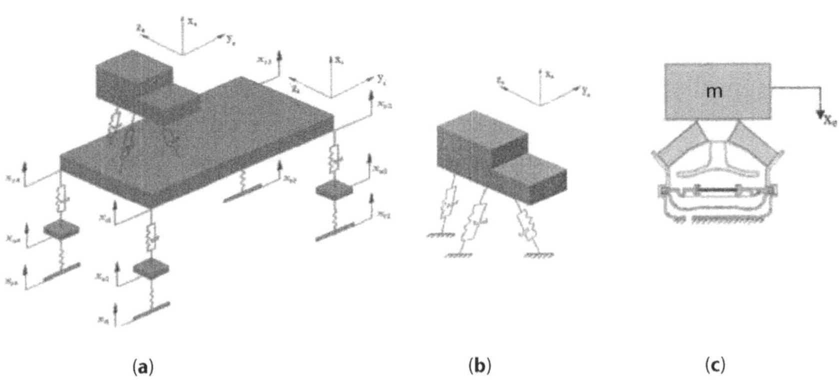

Figure 7.6 Optimization models.

concluded that instead of the complex full-vehicle model, which is too hard to be thoroughly modeled, a model of a 1 degree of freedom (DOF) body mounted on the ground via one HEM can be used. Researchers used a 13 DOF model, 6 DOF model, and 1 DOF model as shown in Figures 7.6a, b, and c, respectively.

Sekulić *et al.* (2013) conducted a fatigue damage assessment of the engine mount bracket using a statistical-based approach. The researchers analyzed two types of strain signals from road loading and compared their effects on the fatigue damage of the engine mount bracket. Strain gauges were attached to the engine mount bracket and were connected to the data acquisition set to capture the actual strain signals when an automobile is driven on two different road surfaces. The strain signals were then analyzed using global signal statistics and the integrated *kurtosis*-based algorithm for *Z-filter (I-kaz)* method. Damage to the engine mount bracket was evaluated using finite element commercial software. The conclusion drawn by the researchers shows that larger strain range amplitude produces more data scatter, higher *I-kaz* coefficient, and larger damage values. The results obtained from the analysis are shown in Figure 7.7, where fatigue damage showed an increment with the respective statistical values of the strain signals.

Su and Liu (2014) used the "*leap frog*" optimization algorithm to minimize the objective function of engine vibration in a mounted 4-cylinder internal combustion engine. They analyzed the mathematical model, for which balancing mass and lead angles were taken as the design variables. The objective function used in this research was the vibratory forces transmitted from the engine to the engine mounts. The objective function was minimized to reduce the vibratory output of the engine.

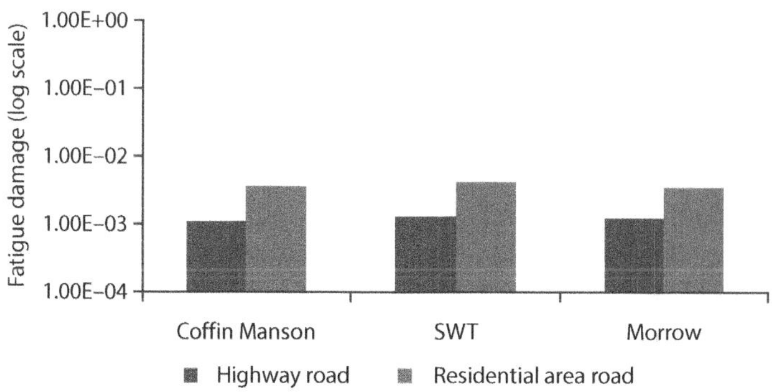

Figure 7.7 Strain signals.

Mahmoudpour *et al.* (2018) presented a simplified method to determine the vibrational amplitude developed by a 4-cylinder engine when supported on visco-elastic mounts. They modeled the engine parts as rigid bodies connected to rubber mounts which were modeled with spring and damping elements. The location, orientation, and stiffness of the mounts can easily be optimized to reduce vibration and noise in the engine design.

Zhang *et al.* (2016) developed a seven degree-of-freedom model for low-frequency engine vibrations that utilizes a two-way coupling assumption. They compared the results of the two-way coupling model with the one-way coupling model. They also identified that the new model properly conserves energy and accounts for gravitational forces.

Gia and Goicolea (2017) conducted an experimental study to determine the rigid body modal content of engine block vibration on a modern heavy-duty inline six-cylinder diesel engine. They used three engine mounts fitted with multi-axis force transducers and utilized standard modal analysis to determine the rigid body modal characteristics and engine mount force signatures of the engine vibration modes of the engine block.

Chalak *et al.* (2014) conducted an experiment on a heavy-duty in-line six-cylinder diesel engine to measure all three orthogonal vibration force components at each of the three engine mounts during standard impact excitation. Modal identification tests on the quiescent engine and during engine operation were conducted.

7.3 Conclusions

This work provides a comprehensive literature review of existing research carried out in terms of design, stress, modal, and optimization analysis of the engine mounting bracket. An effort has been made to comprise all the important contributions to this area and highlighting the most pertinent literature available for investigating the engine mounting bracket. The concluding remarks from the current review are as follows:

a) From the review of available research on engine mounting brackets, it is apparent that most of the research conducted is purely simulation-based on Finite Element Method for economic consideration and validated through experimentation.
b) Out of all the simulation techniques discussed in the present article, the ANSYS simulation package is most widely used

by researchers due to its ease of use and accurate results for assessing the strength of the engine mounting bracket. The same is also applied for shape optimization and material selection of engine mounting bracket.

c) The review shows that engine mounting brackets are designed using different types of materials through simulation to determine the most suitable material.
d) Simulation through COMSOL Multiphysics and mathematical modeling also provides results similar to conventional FEA tools.

References

Abbasi, A., Nazari, F., Nataraj, C., Adaptive modeling of vibrations and structural fatigue for analyzing crack propagation in a rotating system. *J. Sound Vib.*, 541, September, 117276, 2022.

Abouelregal, A.E. and Alesemi, M., Vibrational analysis of viscous thin beams stressed by laser mechanical load using a heat transfer model with a fractional Atangana-Baleanu operator. *Case Stud. Therm. Eng.*, 34, December 2021, 102028, 2022.

Bertini, L., Bucchi, F., Monelli, B.D., Neri, P., Development of a simplified model for the vibration analysis of lawn mowers. *Procedia Struct. Integr.*, 8, 509–516, 2018.

Chai, Q., Zeng, J., Ma, H., Li, K., Han, Q., A dynamic modeling approach for nonlinear vibration analysis of the L-type pipeline system with clamps. *Chin. J. Aeronaut.*, 33, 12, 3253–3265, 2020.

Chalak, H.D., Chakrabarti, A., Sheikh, A.H., Iqbal, M.A., C0 FE model based on HOZT for the analysis of laminated soft core skew sandwich plates: Bending and vibration. *Appl. Math. Modell.*, 38, 4, 1211–1223, 2014.

Chen, J., Zhang, Y., Zheng, Z., Liu, Z., Huang, L., Dynamic analysis of top-excited vibration assisted stuck release in vertical borehole with fatigue-like model. *J. Pet. Sci. Eng.*, 185, November 2019, 106651, 2020.

Demirel, G., II and Kayran, A., Implementation of Dirlik's damage model for the vibration fatigue analysis. *Procedia Struct. Integr.*, 21, 101–111, 2019.

Galván, J.C., Padrón, L.A., Aznárez, J.J., Maeso, O., Boundary element model for the analysis of the dynamic response of the Soria arch dam and experimental validation from ambient vibration tests. *Eng. Anal. Boundary Elem.*, 144, August, 67–80, 2022.

Gao, Y., Sun, R., Leopold, J., Analysis of cutting stability in vibration assisted machining using an analytical predictive force model. *Procedia CIRP*, 31, 515–520, 2015.

Gia, K.N. and Goicolea, J.M., Vibration analysis of short skew bridges due to railway traffic using analytical and simplified models. *Procedia Eng.*, 199, 3039–3046, 2017.

Jabaraj, D.J. and Jaafar, M.S., Vibration analysis of circular membrane model of alveolar wall in examining ultrasound-induced lung hemorrhage. *J. Med. Ultrasound*, 21, 2, 81–91, 2013.

Khazaee, M., Rezania, A., Rosendahl, L., Piezoelectric resonator design and analysis from stochastic car vibration using an experimentally validated finite element with viscous-structural damping model. *Sustain. Energy Technol. Assess.*, 52, PC, 102228, 2022.

Kim, K., Kim, S., Sok, K., Pak, C., Han, K., A modeling method for vibration analysis of cracked beam with arbitrary boundary condition. *J. Ocean. Eng. Sci.*, 3, 4, 367–381, 2018.

Mahmoudpour, E., Hosseini-Hashemi, S.H., Faghidian, S.A., Nonlinear vibration analysis of FG nano-beams resting on elastic foundation in thermal environment using stress-driven nonlocal integral model. *Appl. Math. Modell.*, 57, 302–315, 2018.

Medolago, A. and Melzi, S., A flexible multi-body model of a surface miner for analyzing the interaction between rock-cutting forces and chassis vibrations. *Int. J. Min. Sci. Technol.*, 31, 3, 365–375, 2021.

Nazemnezhad, R., Zare, M., Hosseini-Hashemi, S., Sandwich plate model of multilayer graphene sheets for considering interlayer shear effect in vibration analysis via molecular dynamics simulations. *Appl. Math. Modell.*, 47, 459–472, 2017.

Okeke, C.P., Thite, A.N., Durodola, J.F., Fellows, N.A., Greenrod, M.T., Modelling of hyperelastic polymers for automotive lamps under random vibration loading with proportional damping for robust fatigue analysis. *Procedia Struct. Integr.*, 13, 1460–1469, 2018.

Ruocco, E., Zhang, H., Wang, C.M., Buckling and vibration analysis of nonlocal axially functionally graded nanobeams based on Hencky-bar chain model. *Appl. Math. Modell.*, 63, 445–463, 2018.

Saidin, S.S., Kudus, S.A., Jamadin, A., Anuar, M.A., Amin, N.M., Ibrahim, Z., Zakaria, A.B., Sugiura, K., Operational modal analysis and finite element model updating of ultra-high-performance concrete bridge based on ambient vibration test. *Case Stud. Constr. Mater.*, 16, May, e01117, 2022.

Sekulić, D., Dedović, V., Rusov, S., Šalinić, S., Obradović, A., Analysis of vibration effects on the comfort of intercity bus users by oscillatory model with ten degrees of freedom. *Appl. Math. Modell.*, 37, 18–19, 8629–8644, 2013.

Serdjuks, D., Kurtenoks, V., Tatarinovs, A., Buka-Vaivade, K., Lapkovskis, V., Mironovs, V., Podkoritovs, A., Topcijs, K., Non-model vibration analysis method for health monitoring of structural joints. *Procedia Struct. Integr.*, 37, C, 555–562, 2021.

Serrano, O., Zaera, R., Fernández-Sáez, J., Ruzzene, M., Generalized continuum model for the analysis of nonlinear vibrations of taut strings with microstructure. *Int. J. Solids Struct.*, 164, 157–167, 2019.

Su, W. and Liu, S., Vibration analysis of periodic cellular solids based on an effective couple-stress continuum model. *Int. J. Solids Struct.*, 51, 14, 2676–2686, 2014.

Tang, Y., Zhen, Y., Fang, B., Nonlinear vibration analysis of a fractional dynamic model for the viscoelastic pipe conveying fluid. *Appl. Math. Modell.*, 56, 123–136, 2018.

Wang, X., Lv, L., Li, S., Pu, H., Liu, Y., Bian, B., Li, D., Longitudinal vibration analysis of sucker rod based on a simplified thermo-solid model. *J. Pet. Sci. Eng.*, 196, November 2019, 107951, 2021.

Wang, Y.Q., Li, H.H., Zhang, Y.F., Zu, J.W., A nonlinear surface-stress-dependent model for vibration analysis of cylindrical nanoscale shells conveying fluid. *Appl. Math. Modell.*, 64, 55–70, 2018.

Yalagi, S., Tonannavar, J., Tonannavar, J., DL-3-Aminoisobutyric acid: vibrational, NBO and AIM analysis of N–H···O bonded-zwitterionic dimer model. *Heliyon*, 5, 6, e01933, 2019.

Zhang, J., Guo, P., Lin, J., Wang, K., A mathematical model for coupled vibration system of road vehicle and coupling effect analysis. *Appl. Math. Modell.*, 40, 2, 1199–1217, 2016.

8

Impact of Condition-Based Maintenance (CBM) and Residual Life Prediction (RLP) on Environmental Issues

Jasvinder Singh[1,2]*, Chandan Deep Singh[2] and Dharmpal Deepak[2]

[1]Lovely Professional University, Phagwara, Punjab, India
[2]Punjabi University, Patiala, Punjab, India

Abstract

The objectives of this chapter are to give a quick overview of condition-based maintenance (CBM), including definitions of key concepts, an outline of its history, current uses, and research problems. It also provides insight into several maintenance procedures and lists the benefits and drawbacks of each from distinct angles. The full overview of CBM that follows includes applications of several technologies and approaches that are currently in use. It concludes with unresolved issues related to the implementation of condition-based maintenance systems. This chapter also provides a synopsis of the CBM literature as well as an analysis of CBM as a maintenance tactic. Automotive, aviation, manufacturing, defense, and other industries all use CBM extensively. To make the systems capable of maintenance intelligence, it involves a number of disciplines, including data mining, artificial intelligence, and statistics. Based on past and present system conditions, these disciplines aid in predicting future outcomes. The measured condition monitoring data cannot precisely explain the residual life of a monitored object, but is expected to stochastically correlate with it. The relationship between the unobservable residual life of a monitored item and the current state of accessible condition monitoring history data is established using the stochastic filtering theory. Through this study, readers will understand that how the broad variety of condition monitoring maintenance methods now employed in industries is beneficial for them and how these maintenance methods affect environmental footprints and how these can be analyzed and rectified.

**Corresponding author*: jas.sliet86@gmail.com

Chandan Deep Singh, Davinder Singh, Kanwal Jit Singh and Harleen Kaur (eds.) Condition-Based Maintenance and Residual Life Prediction, (131–170) © 2025 Scrivener Publishing LLC

***Keywords*:** Condition-based maintenance, maintenance, filtering prediction, residual life, data mining, artificial intelligence

8.1 Introduction

In condition-based maintenance (CBM), the actual state of an asset is monitored to determine what maintenance needs to be performed. According to CBM, maintenance should only be carried out when specific indicators indicate a decline in performance or impending failure. Non-invasive measures, visual examination, performance data, and planned tests can all be used to check a machine for these indicators. Then, condition data may be acquired constantly or at predetermined intervals (as is done when a machine has internal sensors). Both mission-critical and non-mission-critical assets can use condition-based maintenance. Contrary to planned maintenance (PM), which is carried out according to predetermined schedule intervals, condition-based maintenance is only carried out when a decline in the equipment's condition has been noticed. Because maintenance is performed only when it is necessary, compared to preventive maintenance, this lengthens the intervals between maintenance fixes.

Several additional components of your maintenance operation must be in place for condition-based maintenance to be effective. This includes implementing a planned maintenance approach that enables you to inspect equipment, identify anomalies, and promptly initiate follow-up repair orders. Check out what AI-powered work order reports can do for you if you want to go one step further and forecast which work orders may result in asset failure. When performance issues are discovered and work is generated, it is also critical to have the proper components and materials on hand. Learn more about predicting your components using artificial intelligence and historical data.

Maintenance has long been a costly and difficult part of supporting the product lifecycle of any particular system Jardine *et al.*, (2005). The frequency of maintenance procedures has often been increased in an effort to reduce failures or fix prior failures of a system (Lorna *et al.*, 2010). Challenges and inefficiencies of the previous generation maintenance system have been addressed by a series of improvement concepts. Preventive and corrective maintenance are the two basic categories into which maintenance procedures are typically categorized. In this chapter, condition-based maintenance (CBM), a third option that some experts may view as a novel concept in maintenance, is discussed. CBM is often included under preventive maintenance.

The Rio Grande Railway Company first presented the concept of CBM in the late 1940s under the name "predictive maintenance." The train firm employed CBM techniques to track variations in temperature and pressure measurements to identify coolant, oil, and fuel leaks in the engine. In terms of minimizing the effects of unforeseen failures and determining when to rectify a leak or replenish a coolant or oil sump, the CBM monitoring techniques were a remarkable success. This concept was very quickly recognized by the US Army, who eventually adopted it as a crucial maintenance tactic for supporting their military equipment.

Throughout the 1950s, 1960s, and early 1970s, CBM ideas and uses began to appear in various industries. The key industries where CBM has been used and has demonstrated various benefits in terms of both efficiency and cost reductions are the automotive, aerospace, military, and manufacturing sectors. The US Department of Defense (army, air force, navy, and marines) and firms like GM, Honda, GE, Digitech, and Honeywell are now investing in and participating with CBM technology applications. By allowing network bandwidth, data collection and retrieval, data analysis, and decision support capabilities for massive data sets of time series data, advancements in information technology have added to and accelerated growth in the CBM technology sector. The focused data collected from a vehicle or other system can provide greater understanding of system performance, health, and failure causes, as well as a prognosis of a system's or subsystem's remaining usable life. For maintaining the mission-critical systems utilized in the aerospace, military, maritime, automotive, industrial, and other industry domains, this is a tremendous advantage. These worthwhile uses and advantages have made CBM a crucial competence area to integrate into a company's product line, whether it be vehicles, aircraft, weaponry, or other items requiring routine maintenance. By incorporating CBM technological enablers into their present and prospective system architectures, these sectors are concentrating on CBM principles and maintenance plans.

The demand for CBM will rise as industries transition to a future where machines are unmanned and closely supervised by humans as they have been in the past. A few system examples that could benefit greatly from CBM ideas and maintenance procedures include robotic systems, unmanned vehicles, windmill systems, manufacturing systems, and oil pumping systems. Businesses that use CBM as a maintenance approach could make considerable financial savings or increase operational efficiency. It might entail a reduction in workforce, a smaller supply footprint, cost savings from second- and third-order failure effects, less downtime, and other advantages relevant to their industry.

8.2 Goals of Condition-Based Maintenance

By gathering and analyzing sporadic or continuous data about the operational status of crucial assets' components, CBM seeks to reduce the overall cost of inspections and repairs (Knapp and Wang, 1992). According to Knapp, Javadpour, and Wang (2000), monitoring, when cost-effective, can give ample notice of impending failures, enabling scheduled repairs based on asset deterioration as opposed to pricey time-based (fixed intervals) repairs or emergency breakdowns. CBM might not always be economical, or the data might not always be there to support its implementation. Monitoring and identifying impending equipment failure is another objective of condition-based maintenance, allowing repair to be proactively scheduled for when it is required and not before. To complete the work before the asset fails or performs below optimal levels, maintenance needs to be triggered by asset conditions far enough in advance of failure.

On the other hand, Raheja, Llinas, and Nagi (2000) stated that a general architecture for CBM is lacking and that present methods for CBM system design are rather specialized. They consequently asserted that each domain area has its own interpretation of CBM, which might not be compatible with the demands of other applications. CBM does not, however, apply to all maintenance assets, and we should only use it in situations when condition monitoring tools are readily available and economically viable (Horner *et al.*, 1997). Reliability and financial maintenance data must be thoroughly analyzed for CBM (Crespo Márquez & Sánchez Herguedas, 2004). Additionally, it necessitates a clear comprehension of asset criticality, failure modes and rates, and potential rewards linked to various maintenance plans. Studies on civil aircraft that described failure patterns and refuted the notion that dependability and operational age are related were mentioned by Sobral (2020). He stated that it was because of this attitude that people began to think that regular maintenance would make an asset more reliable. According to Moubray (2001), time-based overhauls do not increase the reliability of complex products unless there is a predominant age-related failure mechanism. He said that it is very likely that an otherwise stable system would experience newborn mortality problems as a result of a time-based overhaul. Time-based preventive maintenance (PM), according to Wood (1999) and Moubray (2001), frequently fails to maximize the service life of each component; in many cases, the components were being replaced with several hours of useful life still left. Resources are wasted because of PM tasks that are based on strict time intervals; as a result, the building or plant is frequently over-maintained (Mann, Saxena, & Knapp, 1995).

According to Bloch and Geitner (2006), 99% of all machine failures are preceded by specific symptoms, circumstances, or signals. Therefore, by keeping an eye on an asset, action can be made before it seriously affects the organization's performance. The assumption of an age-related failure mode is thus replaced by condition-based monitoring. Additionally, managers might concentrate on just-in-time (JIT) replacement with condition-based monitoring. According to Wood, building automation systems (BAS) or intelligent buildings make JIT replacement easier and maximize the life of each component.

8.3 Maintenance Strategies

To improve the effectiveness of the maintenance function, Coetzee (1999) observed that maintenance management frequently employs some highly publicized strategies. These strategies include reliability-centered maintenance (RCM), total productive maintenance (TPM), condition-based maintenance (CBM), computerized maintenance management systems (CMMS), and auditing systems. However, majority of managers are unaware of the ideal maintenance optimization regime. As a result, majority of efforts are unsuccessful.

Intervention tactics are both a technical and an artistic endeavor (Heckscher, Eisenstat, & Rice, 1994). Hecksher *et al.* claimed that best practice strategies collapse after about a year. They described this phenomenon as a successful failure since there is a lack of knowledge among internal staff members about what they are doing when they begin an implementation process (Heckscher, *et al.*). Coetzee (1999) called for a comprehensive strategy to incorporate CBM within the maintenance role as a result. A comprehensive strategy would consider the building structures and other maintenance assets. According to Shonet's 2003 argument, performance indicators must be created for building systems and components in order for CBM to be implemented effectively. He stated that three factors should be considered when evaluating a building's components:

1. the physical performance of the building's systems,
2. the frequency of system failures, and
3. the actual preventive maintenance (PM) performed on the systems.

The implementation of a balanced scorecard process, the evaluation of breakdown (BD) maintenance, where the run to failure philosophy is prevalent; the evaluation of planned schedule maintenance, where planning and scheduling methodologies are rigorously applied; and the evaluation of CBM, where critical and expensive assets are managed, are all examples of performance measurement systems that should be included in a holistic approach to achieving strategic goals (Amari and McLaughlin, 2004). System parameters gradually deteriorate during the degradation process. The use of CBM necessitates the use of specific analytical tools, such as failure mode, effect, and criticality analysis (FMECA) and RCM to assess the likelihood of failure and how failure would manifest itself, as well as a trustworthy data tool (i.e., CMMS) to record and monitor repairs and associated costs of the assets under consideration.

RCM is a method for choosing the most appropriate and cost-effective maintenance activity for asset elements, and FMECA is a systematic procedure for identifying all the ways in which a failure of an element can occur (El-Haram & Horner, 2002). El-Haram and Horner pointed out that the RCM process assesses each failure to ascertain whether it has any effects and each consequence to ascertain suitable and cost-effective maintenance jobs.

Failure occurs when a resource, part, or element fails to do its task. But in many cases, soft faults-symptoms such as excessive vibration, unusual heat (Knapp *et al.*, 2000), and solid substances in oil occur before complete failures. Asset operators can be alerted and warned of the possibility of an impending failure via warnings and alarms sent by sensors that can detect pre-failure signals. Other assets may fail (hard faults) suddenly and without warning; often, these failures are exogenously created. The real preventive action (repair) is carried out at the level of impending failure with the help of the non-intrusive CBM approach. Tsang observed that standby units, safety measures, or infrequently used assets usually had concealed problems. When an item's correct function is required, hidden failures become apparent.

In light of the possibility of hidden failures at start-up, a CBM regime on standby units, such as protection devices, or infrequently used assets would not be effective. However, Tsang (1995) suggested fault-finding (FF) tasks to address hidden problems. FF tasks are run on assets or items with dormant functionality on a scheduled basis to monitor their condition. An emergency generator that is often not in use until it is a good illustration. Sherwin and Al-Najjar (1999) claim that inspection and condition monitoring are not the best overall maintenance practices when downtime is very expensive unless the asset's or component's combined age to failure distribution has a significant amount of variance, in which case the standard deviation is higher than or (nearly) equal to the mean.

8.4 Determination of CBM Failure Point

Reliable data and efficient algorithms are needed to identify the start of a failure mode and estimate the amount of time until total breakdown. In addition, a disciplined maintenance organization is needed to decide on the proper maintenance response to CBS signals or alarms.

Many maintenance companies with CBM do not react to alerts and alarms in a timely manner. However, when signals from a monitored point cause alerts or alarms, some CBM systems are connected into the CMMS such that a work order is automatically issued. This is especially appropriate for continuous process facilities where downtime and failure can be very expensive.

Moubray and Anderson (2001) discussed the range of condition-monitoring techniques that have been developed for CBM is surveyed. A brief discussion on those which are commonly used to monitor the condition of mechanical systems follows.

8.4.1 Vibration Monitoring

In systems containing rotating or reciprocating parts, such as bearings, gear boxes, shafts, pumps, motors, engines, and turbines, vibration monitoring techniques can be utilized to detect fatigue, wear, imbalance, misalignment, loosened assemblies, turbulence, etc. When such mechanical systems function, energy is released in the form of vibrations with frequency components that may be linked to certain system components. Unless the operational dynamics of the system change, the amplitude of each individual vibration component will stay the same.

Three factors, amplitude, velocity, and acceleration, can be used to describe vibration. The frequency of the vibration affects how sensitive the sensors are that are used to measure these factors. Utilizing amplitude sensors for low frequency signals, velocity sensors for intermediate frequency ranges, and accelerometers for higher frequency signals is the general selection rule.

One method of vibration monitoring involves taking readings from certain locations on a machine of the total vibration energy between 10 and 10,000 Hz. These numbers are contrasted with initial readings obtained from a brand-new equipment. The baseline data serve as the foundation for setting alarm limits. When a reading goes above the alarm threshold, a fault diagnosis will be initiated. As an alternative, to assess the relative state of the equipment, vibration readings are compared to vibration severity charts. Broadband vibration trending is the name of this method, which just analyzes the general machine conditions. The typical microprocessor-based instrumentation for this method tracks the vibration's root-mean-square (RMS) intensity.

The narrowband trending technique measures the total energy over a range of vibration frequencies to track the health of particular machine parts or failure modes. Vibration data in the time domain are captured by the process of scanning vibration signals across a bandwidth. Such information can be converted to the frequency domain to measure the vibration at each frequency component. The machine's vibration signature is the frequency plot that gives a visual representation of each frequency component it produces. A waterfall plot of the machine is created when the vibration signatures of a machine at various times are grouped chronologically and displayed in a cascading fashion on a three-dimensional plot. By observing how the vibration signatures have evolved over time, anomalies in the machine's state can be quickly identified. The instructions for guaranteeing that a vibration monitoring program will not set off false alarms and will give the earliest possible warning of the largest range of machine defects are provided in the work of Serridge (1991).

8.4.2 Process-Parameter Monitoring

A wide range of measurements are included in process parameters, including process effectiveness, heat loss, machine temperature, motor current, and fluid pressure. To track system performance, data on these parameters are typically collected as part of operational procedures. These data have the potential to be used as indications of the system's health in the future. This method of condition monitoring can also be used on non-mechanical plants such as boilers, heat exchangers, filtration systems, and pipework.

8.4.3 Thermography

Thermography uses equipment made to measure infrared energy emissions to ascertain how well plant machinery is performing. Thermal anomalies, such as temperatures that are higher or lower than they should be, are interpreted as warning signs of potential systemic issues. Thermographic methods are best suited to find issues in systems that depend on heat transmission or retention. The surface temperature of a machine surface can be measured using infrared thermometers at a single location. They can be used to spot-check process piping systems and to track the temperature of crucial plant machinery components like bearing caps and motor windings. Infrared imaging will be the appropriate technique when the infrared emission profile of a big area needs to be scanned quickly. The imaging system works similar to a video camera, and the instrument's optics allow users to see the thermal profile of the area being monitored. The number of airborne particles has a big impact on how infrared emissions is measured since ambient circumstances might change. As a result, considerable effort must be made to capture thermal data while adjusting for the impact of such circumstances.

8.4.4 Tribology

The study of the interface between sliding surfaces is known as tribology. Condition-based maintenance employs three tribology methodologies. They are ferrography, wear particle analysis, and lubricating oil analysis. Due to the analysis requiring the use of laboratory equipment such as a spectrometer and scanning electron microscope, these methods are comparatively sluggish and expensive (SEM). In lubricating oil analysis, samples of lubricating, hydraulic, and dielectric oils are analyzed regularly to see if they still have the ability to fulfill the lubricating demands of the application. To maintain satisfactory system operation, the oil will be replenished when its condition deteriorates to an unacceptable level. The findings of the analysis can also serve as the foundation for deciding to switch to a different kind of oil to boost output or cut down on variety. Spectrographic methods are used in lubricating oil analysis to examine the elements present in the oil sample. However, to pinpoint the precise failure mode that could have contributed to the observed deterioration in the oil state, it must be combined with additional diagnostic techniques.

Analyzing wear particles gives precise information on how the machine is worn. This knowledge comes from research on particle size, shape, composition, and amount. Wear particle analysis can be divided into two categories. In the first kind of study, the machine lubricant's solids content, that is, the quantity, size, and composition of the particulate matter in the lubricating oil, is routinely checked to see if it fluctuates over time. To determine the type of wear (rubbing wear, cutting wear, rolling fatigue, combined rolling and sliding wear, or severe sliding wear) that has left its traces in the sample, particle materials in each lubricating oil sample are analyzed using the second method.

Their application calls for specific lubrication. To maintain satisfactory system operation, the oil will be replenished when its condition deteriorates to an unacceptable level. The findings of the analysis can also serve as the foundation for deciding to switch to a different kind of oil to boost output or cut down on variety. Spectrographic methods are used in lubricating oil analysis to examine the elements present in the oil sample. However, to pinpoint the precise failure mode that could have contributed to the observed deterioration in the oil state, it must be combined with additional diagnostic techniques.

8.4.5 Visual Examination

Visual examination of plant machinery provides a quick way to find issues including leaks, loose mountings, and structural cracks. The maintenance team can conduct this examination while gathering the regular data required by the other condition-monitoring methods. As a result, performing a visual inspection typically has little added expense.

8.5 Decision-Making in Condition-Based Maintenance

There are three types of decisions that need to be made in condition-based maintenance:

1. Selecting the parameters to be monitored;
2. Determining the inspection frequency; and
3. Establishing the warning limit (the trigger).

The sort of plant facilities that need to be covered, the accessibility of dependable surveillance equipment, investments in instrumentation and training, the amount of required staff, and running costs are only a few of the variables that influence what should be monitored. It should be noted that an impending functional failure may manifest itself in various ways, each of which may be discernible at various stages of the unit's degeneration. For instance, altered vibration characteristics can serve as the first indication of a bearing failure. Increased particle content in lubricating oil samples, audible noise emission, and heat build-up on the bearing caps follow in that order. Symptoms that appear in later stages give the occurrence of functional failure less time to prepare. When deciding on the most economical CBM process, it is important to take into account all of these symptoms and the condition-monitoring strategies that apply to them.

Determine the time between inspections if equipment condition is not continually monitored so that timely action can be taken to either prevent functional failures or to lessen the effects of those that cannot be avoided. Over-inspection will occur if this interval is too short. Too much time will increase the possibility of unanticipated outages. The time interval (T) between the earliest instant at which signs of an imminent failure become detectable and the instant at which a catastrophic failure occurs is a method for determining inspection intervals. The sum of the anticipated preparation time for any required preventive measure and the scheduled inspection interval should not be greater than half of T. The decision-making process is complicated by the uncertainty related to the estimation of 'T' that is present in many real-world circumstances. The following sections of this paper will introduce some of the decision models created to meet this type of inspection difficulty.

Both static and dynamic warning limitations may be used in a condition-based maintenance program as catalysts for replacement and diagnostic action. The predetermined thresholds of the measured data are known as static limits. Examples of static limits for vibration monitoring include the ISO vibration severity charts. Even though CBM activities that are triggered by static limitations are simpler to manage, they lack the diagnostic capability to foretell when the alert will be met. On the other hand, dynamic limits are employed to track the pace at which the measured parameter changes. When using dynamic warning limits, a CBM procedure prioritizes the rate of change of the monitored parameters over their actual value. Using dynamic limitations will undoubtedly improve a CBM procedure's automatic diagnostic capability.

8.6 Decision Models for CBM

The topic of inspection is one of the main classes of decision problems in CBM. It deals with circumstances when inspection duties are carried out on equipment at regular intervals to ascertain its actual condition. The outcomes of each examination will determine one of two things:

First, decide whether to replace, repair, or leave the system in its current condition. Second, decide when the next examination should occur.

Valdez-Flores and Feldman (1989) surveyed the decision models for the inspection problem of single-unit systems. Some of these models are briefly introduced below.

Luss (1976) proposed a model with these assumptions:

- perfect inspection;
- opportunities for inspection occur every Dt time units;
- the system's level of deterioration has a discrete state space, and the holding time in the various states is exponentially distributed;
- infinite planning horizon;
- a failure can only be detected by an inspection;
- occupancy costs are non-decreasing with degrees of deterioration;
- repair costs are the same for all states except failure; and
- repair at failure is more costly than repair immediately after an inspection.

A control limit policy with state-dependent inspection interval is developed to minimize the expected cost per unit of time. The decision variables of the model are

- the critical state triggering a preventive repair and
- the time to the next inspection ti where i is the observed state at inspection.

The necessary condition for the existence of an optimal solution and an iterative procedure to solve the model are provided.

Sengupta (1980) offered a development of Luss's concept. In this extension, the repair cost is a function of deterioration that increases with time, and deferred repair or replacement is permitted. The answer is also shown, along with an iterative algorithm.

To deal with similar issues, Zuckerman (1980) developed a model: perfect inspection; fixed inspection intervals; shocks occur in an HPP and the degrees of damage caused by subsequent shocks are IID on; infinite planning horizon; a failure is discovered only at inspection; and the repair and operating costs are non-decreasing with degrees of deterioration. The presentation of a control limit policy for reducing predicted cost per unit of time. In this paradigm, the critical state that demands a preventive repair serves as the decision variable. It also specifies the prerequisite for the existence of an ideal solution. Although a method for solving the model is not provided, it is noted that the difficulty depends on the layout of the system's survival function and the distribution of the damage brought on by the shock.

With these conditions, Rosenfield (1976) considered a somewhat different scenario: Perfect inspection; discrete-time Markov system deterioration; operator's three options of repair, inaction, or inspection; and non-decreasing repair and operation expenses with system deterioration. The target function is minimizing long-term predicted cost per unit of time, or anticipated total discounted costs. Also provided are the system parameter requirements under which a four-region strategy would function best.

Kander (1978) considered the following three inspection policies: pure checking, shortened checking, and checking plus monitoring. In these rules, discrete deterioration levels are assumed, and the next inspection time is chosen based on the system's most recent state. The next checking is always scheduled under the pure checking policy when the item is discovered to be in an undefeated state. According to the shortened checking policy, the item will be replaced after it has degraded to one of its late phases. Except for starting a continuous monitoring operation when the item is discovered in a late stage, the third policy is similar to the truncated checking policy. These models' main goal is to reduce predicted long-term costs per unit of time. There is no description of how to find the best answer.

Other varieties of inspection models have also been created to deal with issues including poor inspections and dangerous inspections where each inspection will result in an instantaneous failure or increase the failure rate, and inspection of a stand-by system.

8.7 Proportional Hazards Modeling

According to the traditional method, an item's failure risk is only based on its age or usage. It is acknowledged that equipment condition monitoring techniques, including vibration analysis and spectrographic analysis of oil samples, also offer a wealth of data that can improve the accuracy of the assessment of such risk. The approach for this evaluation is provided by proportional hazards modeling (PHM), a multivariate regression analysis technique first presented in the work of Cox (1972). The PHM technique was first mostly used in the medical industry. The open literature first documented attempts to apply the approach to evaluate the reliability of airplane, marine, and locomotive engines.

The condition-based maintenance decision process can benefit from the additional data collected from condition monitoring jobs and analyzed by the PHM technique. To maximize replacement decisions, Makis and Jardine (1992) offer a control theory technique that includes PHM risk assessment in the cost consideration. The PHM procedure is used to assess the level of deterioration in that method, and the transition between levels is discretized and modeled as a semi-Markov process. Inspections take place at regular periods. At each inspection, a choice is made between replacing the item right away and planning to keep using it for another x inspection intervals. Both the state and the age affect the value of x.

8.8 Maintenance Planning and Scheduling

Planning is the process of deciding what decisions and activities will be taken in the future to achieve the desired goals and aims. Setting up future actions in advance aids in reaching objectives as quickly and effectively as possible. It reduces expenses, risks, and lost opportunities. Additionally, it may strengthen the organization's advantage over rivals. Depending on the planning horizon, the planning process can be divided into three fundamental levels, where three types of planning exist:

- long-term (year-based),
- medium-term (monthly-based), and
- short-term (daily and weekly plans).

Different decision levels, such as strategic or tactical, need planning. Different organizational levels, including corporate, business, functional, and operational, may carry it out. Strategic decisions affect the essence of the business's existence as a corporate entity, whereas tactical decisions have an impact on how the business is run at a certain point in its growth trajectory. Strategic planning establishes the organization's long-term goals and plots a course for achieving them. The tactical level of long- and short-term planning focuses on choosing paths within a predetermined strategy for accomplishing long-, medium-, and short-term goals and targets. Strategic planning can be carried out at the functional, business, or corporate level and is, by definition, a long-term strategy. However, long-term planning need not always be strategic. Planning often entails deciding on the activities or tasks to be accomplished as well as the resources required to carry them out, regardless of its nature and goal.

Putting the tasks outlined in the plan into a time frame is the process of scheduling. The intended goals, the relationships between the many scheduled tasks, the availability of resources throughout time, as well as any other internal and external limitations and constraints, are all taken into account. The effectiveness of the resulting schedule is typically assessed using a performance metric with respect to the task or tasks' intended aim. By meeting deadlines, completing tasks on time, or making good use of resources, performance metrics can be connected to various expenses. When used in this context, the term "maintenance" refers to all actions taken to ensure that the system's components are reliable, available, and capable of performing at a standard level of quality. It comprises tasks linked to managing risk, managing human resources, and keeping an inventory of replacement parts. In a larger sense, it refers to all choices made by the organization at all levels regarding the acquisition and upkeep of high levels of availability and dependability of its assets.

In majority of organizations and systems, including those in construction, manufacturing, and transportation, maintenance is turning into a crucial functional area. It is becoming into a significant functional area in all kinds of businesses, influencing and being influenced by many other functional areas like production, quality, inventory, marketing, and human resources. It is also starting to be seen as a crucial link in the global supply chain for business. The high cost of maintenance, which is believed to account for about 30% of the overall operating expenses of contemporary manufacturing and construction firms, is reflected in its increasing rule. Visser presented a system view of a maintenance system that places maintenance in relation to the enterprise system, as depicted in Figure 8.1.

All forms of corporate business decisions that include potential future large investments should take maintenance into account, whether they are tactical or strategic, long-term or short-term. For instance, choosing to buy a new facility with poor maintainability could be a complete disaster for the entire company. The ability to maintain the plant and the capacity to do so should be considered while planning its capacity.

The most crucial components of effective maintenance management are planning and scheduling. Reducing maintenance costs, delays, and disruptions, as well as enhancing, are all greatly influenced by effective planning and scheduling.

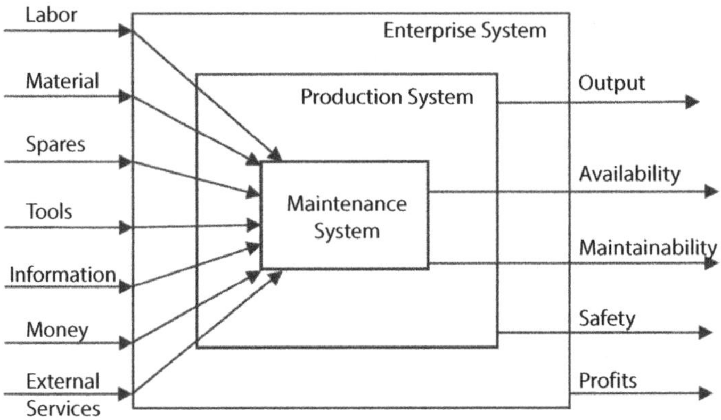

Figure 8.1 Input–output model of the enterprise.

8.9 Maintenance Concepts and Strategies

8.9.1 Reliability-Centered Maintenance (RCM)

By choosing the ideal blend of corrective maintenance, schedule-based (or preventive maintenance), and CBM to completely support the reliability of the system in any given operational environment, RCM enables the design of the maintenance strategy. Which maintenance strategy should be used, and to what extent? These are some essential questions that RCM provides organized solutions to. What justifications are there for the maintenance plan? Is it possible to lower the expense of maintenance? Is the maintenance approach feasible? (In terms of efficiency and in terms of operations and costs). According to Moubary (1997), RCM is a procedure used to decide what needs to be done to ensure that any physical asset continues to function as intended by its users within its current operating environment.

RCM is considered while developing a system for supportability, and the maintenance plan is further developed throughout the systems engineering process (i.e., the validation, verification, and test phases) and through the system's initial fielding. The RCM model is improved by system testing and field feedback on system failures. It is in this feedback loop that CBM technologies can have the most impact on enhancing system performance throughout the course of a system's lifecycle. It makes it possible for businesses to use a process of continuous improvement to outperform rivals. A procedure or action known as corrective maintenance (Neelamkavil, 2010) or the run-to-failure model is necessary to stop a failure that has already happened or is now happening. To return the system to the way it was when it was brand-new, it can entail repairs, component replacements, or other actions. Regardless of the state of the system or subsystem, the schedule-based maintenance or preventive maintenance model (Chen and Trivedi, 2001) is a maintenance philosophy based on predetermined intervals. For instance, even when a sizable percentage of the oil's life is still left, changing the oil in an automobile is routinely done either depending on miles or time interval.

The concept of preventive maintenance (Neelamkavil, 2010) was developed to decide how much maintenance should be allocated for a particular system or subsystem. CBM, however, has been developed from RCM. RCM addresses functional safety, environmental, and operational failure modes while concentrating on the operating environment of the system. It carries out a thorough study of the dependability of the system's individual components and the system as a whole in an effort to identify the primary cause of the failure(s) in question. It is common knowledge that if a component or system is maintained and controlled to operate within the toler ance limits of normal operation, it may last longer without compromising performance.

A strictly preventive or corrective maintenance strategy results in inefficient human utilization, system downtime, income loss, and other waste. For instance, most vehicles just change the oil periodically rather than according to how well it is doing at a particular moment in time. Without endangering the engine, the rejected oil might be used even more times. It could not be the best use of the available resources. Most machines and plants in manufacturing facilities receive routine maintenance to prevent unplanned maintenance and downtime (Kozusko, 1986).

For a specific facility, this maintenance method results in recurring downtime, which causes production lulls and inefficiencies. In a sense, it ensures the plant will run well, but on the other hand, it has an impact on the productivity of the plant in terms of cost, downtime, labor engagement, etc. Even though it might be costly and ineffective, this type of maintenance is common and is used by most sectors.

CBM, on the other hand, focuses on the system or subsystems' current state. To minimize downtime at the most advantageous period, the underlying maintenance process eventually initiates a business activity (such as a supply or maintenance action). It suggests that a system is used to the farthest extent possible given its expected performance level. Before it falls short of predetermined performance standards, it is then replaced or repaired. As a result, it allows the system to continue operating as long as it is performing within predefined performance limits.

To "lean" out their maintenance procedures and practices, the military, manufacturing, food processing, aerospace, automotive, and several other industries have been contemplating the usage of CBM applications (Neelamkavil, 2010; Williams *et al.*, 1994; Barajas and Srinivasa, 2008). This strategy is frequently more effective and valuable than the ones currently in use. When CBM is incorporated into your maintenance philosophy, the emphasis is placed on the component's or system's current state rather than schedule-based or run-to-failure types of maintenance procedures. Not all subsystems are suitable for monitoring and failure prediction. Therefore, as part of a system's maintenance strategy, one must select the right balance of corrective, preventive, and CBM depending on the system or subsystem in question and the results of the RCM analysis.

8.9.2 Equipment Failure Behavior

Developing a CBM approach requires a thorough understanding of the equipment failure behavior. The P–F curve is a common name for the curve that describes the failure's behavior. The P–F curve, shown in Figure 8.2, illustrates the linked parameter's fault behavior. The fundamental presumption is that the parameter has a tendency toward the maxim "greater is better." This graph illustrates the points at which failures begin to occur but are not yet observable (P), at which they become detectable (P1) and are often referred to as prospective failures, and at which the system fails and is referred to as functional failure (F) point. Many techniques can be created to identify these intermediate states, P2–Pn, at many additional places between P1 and F. P–F interval is the amount of time it takes for a prospective failure to develop into a functional failure. This interval is the only time frame for all required analyses and maintenance. This interval's order is determined by the properties of a specific system. Other probabilistic measurements, such as failure detection threshold (FDT), which is the proportion of a component's life that must pass before an approaching failure may be identified, are used to identify when system failures may be detectable. For instance, if the FDT is 0.7, a failure is only likely to develop or be noticed within the first 30% of a system's lifespan. The FDT differs from system to system since it is highly dependent on system features.

8.9.3 Condition-Based Maintenance (CBM)

CBM (Nickerson and Hall, 1995) involves monitoring the condition of mission critical and safety-critical parts in carrying out maintenance

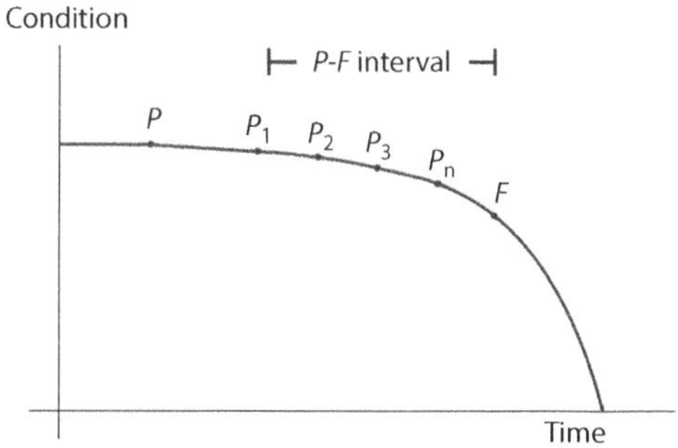

Figure 8.2 Equipment failure behavior (P–F curve).

whenever necessary to avoid hazards rather than following a fixed schedule. According to the US Air Force, CBM is this:

A collection of upkeep procedures and skills known as condition-based maintenance (CBM) are developed from real-time evaluations of weapon system health received through embedded sensors and/or external tests and measurements made with the aid of portable equipment. CBM aims to only carry out maintenance when it is clearly necessary.

Engine oil quality detection is one application for CBM. A CBM system has been implemented by some automakers, including GM, to identify the oil quality based on the longevity of oil components. Based on the present oil's viscosity, total acid number (TAN), total base number (TBN), and other characteristics, this system begins informing operators of the remaining useable miles. In this instance, a system like this enables the oil to be used to the fullest extent possible as opposed to being prematurely replaced at set intervals. The academic community and a range of enterprises have examined CBM strategies from different perspectives. Later sections of this study concentrate on the various CBM components and facilitators.

8.9.4 Condition-Based Maintenance Plus (CBMp)

The regular CBM components are subject to RCM examination in the CBM. The following is the US Air Force's definition of CBM: These fundamental ideas are expanded upon by condition- based maintenance plus (CBMp), which includes additional technologies, methods, and techniques that enable better maintenance and logistical operations. The planning, acquisition, maintenance, and disposal of a weapon system's capabilities will consider these present and upcoming technologies, procedures, and capabilities. This suggests that CBM is more than just a "box" you can purchase and connect onto your platform or system; rather, it is a collection of integrated technologies, procedures, and capabilities that work together to make CBM a reality.

8.10 Condition-Based Maintenance (CBM) Technology Enablers

8.10.1 Diagnostics

Finding the fault in a system after or during the defect's occurrence is known as diagnostics. For instance, a mechanic for a vehicle might use a computerized read-out instrument to try and locate a fault code for a malfunctioning vehicle, and then compare the code to a database of faults to get information on the related defect. According to Barajas and Srinivasa (2008), there are several ways to examine the diagnostic process of any system, including ad hoc system and machine-level analysis, vibration analysis, lubrication and oil analysis, infrared thermography, ultrasound, high-speed video, and fibre optic. System diagnostics is, for the most part, a reactive method of locating system flaws.

8.10.2 Prognostics

Prognostics is the process of foretelling a system's failure in the future by examining its past and present operating conditions, or by tracking how frequently those conditions deviate from normal. There are two main difficulties with prediction:

1. time to failure prediction or remaining useful life prediction and
2. trust value estimation (TVE).

The authors' V diagram of the CBM process is depicted in Figure 8.3. The techniques for gathering data and populating a database are displayed on the left side of the V. The "expert system analysis," which forms the vertices of the V diagram, is the core of the CBM capacity. The prognostication and display operations are described on the right side of the V diagram.

A real physical system receives real-time input from the system (different parameters like the outside temperature and required speed) and outputs a set of results (actual speed, actual system temperature, etc.).

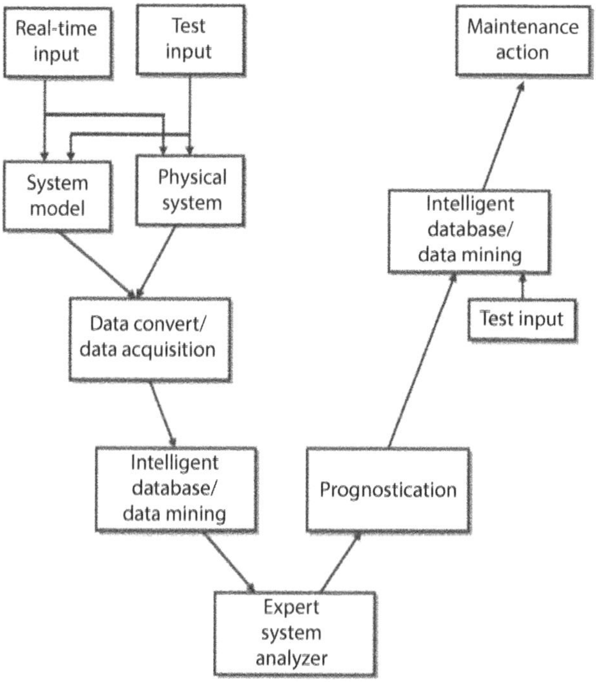

Figure 8.3 The V-architecture of CBM.

To determine whether the real system operates as intended and generates the appropriate outputs, test inputs are fed into both the real system and the system model. For each of the test inputs, the system model generates the anticipated output. Sensors, wired and wireless techniques, as well as other data collecting methods are used to gather the outputs of the system model and physical system, which are then stored in a database. An expert system is used to compare and analyze the data. From the historical data that have been stored, the intelligent data mining stage retrieves the useful data features. The prognostication block uses a variety of methods to determine how long system components are still functional before needing maintenance, then it delivers the results to the display unit and maintenance action unit. The prognostic modeling unit, which verifies the prognosis model using a set of test inputs, also feeds the display unit.

8.10.3 Usage-Based Modeling

The remaining useful life of components can also be predicted using usage-based models. One such example is the oil life prediction system used by General Motors. It is based on driver activity, temperature peaks, and engine revolutions (hard accelerations, excessive use of breaks, etc.). Extensive testing and analysis are needed for these usage model types. If in-depth research is not done beforehand, there is a significant margin of error and the forecast is only as accurate as the model. Therefore, while creating and implementing usage-based models, model validation and verification testing are of the utmost importance.

8.10.4 Data Mining in CBM

One of the important enablers in the construction of a CBM capacity is the use of data mining (Hastie *et al.*, 2009) and management (Tsang *et al.*, 2006) methodologies. In general, all systems, including those in manufacturing, the auto industry, and the oil and gas industry, occasionally dump significant amounts of data from many sources. In general, the amount of data gathered by the sensors can be enormous and not all of it may be useful or significant. Building the model and obtaining useful data are made easier by data mining. Finding the relationship that expresses how the change in one or a set of variables affects the other variables is known as model building. Data preparation, model generation, model validation, and deployment are the four processes that typically make up the data mining process.

Data cleaning, data transformation, clustering of similar types of data, data integration, and data reduction are all part of the data preparation process (Zhang *et al.*, 2003). The raw data that were directly obtained from the system that was being operated or tested was preprocessed at this stage. In general, raw data may be lacking some desirable properties or be noisy (i.e., errors or inconsistent that may cause irregularities in the pattern). According to Pyle (1999), the preparation stage entails a number of tasks, including auditing the data, upgrading and enriching the data, identifying sample bias, figuring out the data structure, creating a ready information environment, surveying the data, and other data quality stages. The data are changed in this stage from raw to meaningful so that the model may be built.

The following step in the data mining process, model creation, is based on the information obtained in the previous stage. At this point, data patterns or characteristics start to take shape. To project system life or failure, trend analysis, and multi-variable correlation are used as a means of prediction. A model is a collection of statistical correlations over a time period based on the predetermined behavior of the operational or experimental system. For a typical course of operation, failure states, and other states between these two, various patterns can be found to obtain full training sets.

The validation of data patterns produced from the prior sets is the focus of the model validation step. It is possible to perform this utilizing various well-known data sets. After that, the models should be verified to make sure they adhere to the prognostic output's purpose.

To anticipate the remaining system life, the generated model is then applied to a new set of data (i.e., prognostics). Data mining language refers to deployment as the last stage. Neural networks, decision trees (CART and CHAID), evolutionary algorithms, closest neighbor search, rule induction, and other notable techniques and algorithm approaches are widely employed in data mining.

8.10.5 Artificial Intelligence in CBM

The creation of robust reasoning algorithms and prediction methods is a component of artificial intelligence (Russell and Norvig, 2002). According to Smith *et al.* (2003), these algorithms play a significant role in the forecast of system life and are one of the crucial elements for the success of CBM. The most well-known uses of AI techniques are in search capabilities, such as the creation of search plans.

The emphasis is on comparing various plans based on their merits and selecting the best order to accomplish the objective. To anticipate the system life based on the real-time attributes gathered from the system, similar methods can be used in CBM applications.

Machine learning algorithms are a key component of AI. These algorithms' objective is to find the intricate pattern in the sensory data obtained from the operation or system-on-test, and then they produce the deft decisions to be made from these data sets. This has a one-to-one relationship with the CBM system since the CBM system's objective is to make decisions based on data collected from the system during the course of the system's life cycle (Prajapati and Ganesan, 2012).

8.10.6 Open System Architecture-CBM (OSA-CBM)

The organization MIMOSA, which is active in the creation of the standards for CBM, created OSA-CBM (MIMOSA, 2006). This group includes the US Army and a number of foreign businesses. An end-to-end CBM system can be realized with the aid of the information flow standard known as OSA-CBM. This was created using unified modeling language, which bridges the gap between several communities for the flow of information.

For the communities, including engineers, scientists, students, and others involved in the creation of CBM systems, OSA-CBM offers the common framework. Advisory generation, prognosis assessment, health assessment, state detection, data manipulation, and data acquisition are its six building blocks. The final advisory state instructs what should be done. It has numerous benefits and crucial characteristics (e.g., cost, specialization, competition, and co-operation). The prognosis stage focuses on estimating anticipated future failures as well as the system's remaining useful life. A health evaluation specifically searches for flaws or failures that are going to occur. As the name implies, the state detection stage uses the data obtained from the previous stage to determine the device's status. The data collected from the device-on-test are analyzed during the data manipulation stage. Data conversion into a format that can be read by humans is known as data acquisition. Each maintenance method has a limitation of its own, and the best one to use depends on the application's needs. CBM is the most effective of them for ongoing maintenance needs. In contrast, the cost of initial infrastructure deployment is higher, but over the course of a system's lifecycle, it offers several advantages. It is a good idea to establish such maintenance concepts for applications such as fleet management, food processing, plant maintenance, and other applications. All industries that need to support systems from a maintenance and supply viewpoint can benefit from CBM (e.g., fleets, plants, and weapon systems).

8.11 Survey of Recent Developments in CBM

Different research communities are currently developing CBM capabilities in various ways. The development of CBM applications for the Joint Strike Fighter is heavily influenced by the US Air Force. When designing and creating new weapon systems, they also intend to consider CBM. "Condition based maintenance plus (CBMp) is the use and integration of knowledge-based capabilities, technologies, and processes to enhance the maintenance effectiveness and dependability of Department of Defense systems and components." To gather data, enable analysis, and assist the decision-making processes for system acquisition, sustainment, and operations, CBM employs a system engineering methodology. It has been discovered that employing a CBM technique is more effective and contributes to cost savings over the course of a particular system. Any degradable system's maintenance plans have always been a source of worry and expense. Numerous initiatives to enhance maintenance procedures have been attempted in the past. Due to its predictive nature and favorable effects on the supply chain and fleet management, attention has recently switched toward CBM. The core of CBM intelligence lies in the fault prognosis prediction algorithms.

In the part that follows, we discussed a few current initiatives to enhance fault prediction and CBM. Yang (2002) investigated the one-step-ahead prediction capabilities of the Kalman filter (KF) for DC motor maintenance. KFs are particularly effective in terms of computation for applications with huge states. According to Yang, the one-step forecast error is acceptable. However, it is difficult and costly to compute and is not suitable for extremely critical systems. Any degradable system can use the reliable and all-encompassing method Grall *et al.* (2002) proposed. However, they constrained the technique with a few restrictions and dependencies that do not seem to be particularly useful in practical applications. Based on the analysis of normal behavior, Ferrell (2000) presented a method for identifying the system's abnormal behavior. To express a machine for one classification of a non-stationary class as well as additional stationary classes, they presented a general support vector representation. By adding new classes, it manages data even when some classes are lacking. This approach makes it easier to forecast how the system will perform in the future.

Lu *et al.* (2007) looked at a predictive CBM capability to forecast the future state of a failing system. Modeled as continuous states, the degradation states depend on random variables to determine fault likelihood. The maintenance cost is the main focus of the suggested technique, and prediction accuracy becomes the most crucial element. CBM is specifically used for critical systems to detect and predict fault occurrences within very short time intervals. A good maintenance system, according to the authors, must strike a balance between prediction accuracy and maintenance cost. Additionally, Rausch (2008) looked into a CBM technique that provides the connection between CBM systems and continuous state/time degradation models. This method is not dynamic, as he would discuss in his subsequent study. This has been the focus of our suggested system.

Chen and Trivedi (2002) gave a closed-form analysis of a system that deviates from its usual course of operation and encounters the failures of poison. To maximize system availability, reduce operating costs, or shorten the mean time to failure given specific system availability limitations, authors looked into a closed-loop strategy to optimize inspection intervals. It was discovered that the system's uptime utilizing a polynomial equation-based solution is deterministic, quicker, and appropriate for real-world applications. Using CBM, Yam *et al.* (2001) created an intelligent decision support system. Based on a recurrent neural network, this system adds capability parameters to forecast potential fault scenarios in the future. Although very sluggish, neural network technologies are effective. These methods also demand a sizable data collection for training, which proves to be a significant barrier for real-time applications.

A good failure prediction method has been described by Angel *et al.* (2000), focusing on accuracy, precision, and confidence. For important or real-time applications, their findings indicate a 5% variation in predicted failures, which is still high. A general framework for diagnostics and prognostics has been proposed by Baruah *et al.* (2006). They have created an online method that eliminates the need to interface with the physical system, hence requiring less labor. The multi-basis clustering and efficient cluster tracking are the foundation of this approach. Due to certain subjectivity bias brought on by human decision inputs, it has some downsides. Due to the fact that both authors described the online maintenance technique, Kwon *et al.* (2009) also suggested a methodology that is comparable to that in the work of Baruah *et al.* (2006). The analysis of online data for remotely placed systems is the main topic of this essay. The robot's harmonic drive is primarily connected to the Internet, and a distant off-line analyzer can access all the real-time data. The breakdown of any robot subsystem can be predicted using a mathematical model. The findings of the trial indicate that this tactic is more effective at boosting system uptime. The various optimization techniques have also been suggested for practical optimization of CBM enabled systems.

Semi-decision Markov's process (SDMP) was employed by Chen and Trivedi (2005) to optimize a CBM strategy. Chen has demonstrated that optimizing for both inspection rates and maintenance policies is preferable to doing so solely for inspection rates. SDMP is a tried-and-true methodology that works well for simulating a variety of failure scenarios. On the other hand, it is not suitable for time-dependent degradations and requires a big data set for training. As a result, SDMP is not quite beneficial for CBM applications, particularly for failures of the extremely severe and time-varying variety. Schmitigal and Moyer (2005) talked about a diesel engine oil condition monitoring system. Using an on-board sensor system, they primarily examined two important difficulties in this study project:

soot accumulation and oil oxidation. To verify the correctness of the on-board sensor system results, the findings from the on-site analysis were compared with the findings from the on-board sensor systems. According to the results, this specific on-board sensor technology might be used to forecast the oil quality straight from the fleet without the assistance of an oil analysis lab.

In the study, Sondalini claimed that aging/degrading causes account for 15%–20% of equipment failures and random failures account for the other 80%–85%. Only by studying failure patterns or applying certain failure prediction models can the random failures be predicted. When failures are this random and caused by other variables besides aging or degradation, schedule-based maintenance is ineffective.

The improvement of a smart wireless internal combustion engine (SWICE) device that has been considered as a mini-vehicle computer system is covered in the work written by Zachos and DeGrant (2009). An at-platform device, like as a rugged laptop, can access the vehicle subsystems via a wireless link using the smart wireless diagnostic sensor (SWDS) that is part of the SWICE. An ad hoc network setup using an off-platform fleet data collection terminal is made easier with the aid of SWDS. The wireless interface of the device can be used to download some usage information such as miles, hours, faults, temperature, pressure, and other vehicle factors, which can then be used to immediately evaluate the vehicle data.

8.12 Application Areas of CBM

CBM is used in a wide range of industries, including manufacturing facilities, the process sector, ground vehicles for the military, navy and air force, IT infrastructure, commercial vehicles, and aviation and aircraft. We gave the IT infrastructure some basic guidance on engine oil, vibration analysis, and temperature analysis. These are the few fields in which CBM is applied. CBM is being used in the highly targeted realm of manufacturing.

8.12.1 Automobiles

The engine oil used in military trucks, commercial trucks, and other engines that consume engine oil are the subject of this section's study. In typical service conditions, it is customary to change the oil every 3,000 or

7,500 miles or at any other interval, depending on the manufacturer. It was covered in earlier parts; however, it is possible that the oil quality did not go below the necessary level. It might continue to serve the engine for a few hundred miles without having an impact on performance. Some modern cars now come with sensors that monitor the oil's condition, alerting the driver to change the oil only when the quality drops below a predetermined level rather than at predetermined intervals. The reasons why oil deteriorates over time can vary. During engine running, engine oil mostly picks up some impurities. By extracting some samples from the engine, it may be examined. These substances, such as antifreeze, fuel, or dust, cause the engine oil quality to be much worse than it should be under regular operation. Here are a few instances of pollutants and their effects:

- silicon dust and fine dirt from a leaking air system;
- chromium-piston ring wear;
- copper, tin-bearing wear, and sodium-coolant leak; and
- iron-liner, crankshaft, camshaft, cylinder liner, and timing gear wear.

The quality of engine oil (Schmitigal and Moyer, 2005) depends on the five main components of the oil as listed:

- viscosity index (VII);
- TAN;
- TBN;
- oxidation induction time; and
- dielectric strength

Oil viscosity a measurement of its thickness; an oil with a lower viscosity flows more readily. However, because of their thickness, heavy oils are more difficult to flow (i.e., highly viscose). These oil grades are defined by society of automotive engineers (SAE) and are denoted by the symbol xWy, where x is the oil's pour point in degrees Celsius, "W" is the winter grade, and the number after the "W" is the oil weight, such as VII at 100°C. A few of the engine oils that are readily available on the market are 0W20, 5W30, and 10W30, which are multi-grade oils appropriate for both low and high temperatures. One of the most important factors in determining the condition of old engine oil is viscosity. Oils of a single grade do not contain VII. Another crucial component of the engine oil quality metric is TAN. This figure represents the oil's overall acidity index and considers all of its available acidic components. Because acidic components such as rust, corrosion, and dust increase with the use of oil, this figure rises as oil quality declines. Depending on the oil, TAN numbers usually range from two to four.

The TBN number is concerned with the alkaline oil constituents that balance the engine oil's acidity. Usually, this number falls between 5 and 15. The TBN number decreases as oil is used since it contains more acidic components. This figure loses its value for high-quality lubricant when it falls below three. Following a brief analysis of a few oil quality parameters, the ideal oil change interval that maximizes engine oil use can be identified.

8.12.2 IT Infrastructure

One of the major problems with an IT infrastructure is maintenance. The information technology revolution has elevated IT infrastructure to the same level of significance as manufacturing plant machinery. Today, hard discs are used to store essential data in electronic form. Large data centers are being used by multinational corporations to store electronic data. The biggest loss for any large company could result from a single point of failure in the servers. These problems raise important challenges about how to maintain infrastructures to deliver high-quality services and lower the likelihood of data loss. Disk failures are caused by many other things. There are four widely used parameters together known as SMART parameters:

- reallocation counts;
- probation counts;
- offline reallocation; and
- scan errors.

Disk drives had 14% higher failure rates following the initial reallocation than those without such faults, according to research by Pinheiro *et al.* (2007). The crucial threshold value for this parameter is one. In comparison to regular disc drives, the disc drives have a 16% higher probability after the first probation count. Additionally, one is the necessary threshold for probation to count. Disk drives are also 21% more likely to fail after initial offline reallocation than regular disc drives. Disk drives that experience their first scan mistake have a failure probability that is 39% higher than those that do not. The aforementioned parameters have all been researched for 60 days. Because disc failures are based not only on SMART parameters but also on some other signals including seek errors, CRC errors, power cycles, calibration retries, spin retries, power-on hours, and vibrations, prediction models based on these SMART parameters may not produce the expected behavior. According to the total analysis, scan errors, reallocation count, offline reallocation, and probation count are all determined to be error-free in about 56% of failures. Even though all factors are added to the failure analysis, with the exception of temperature, 36% of failures still showed other signals as the cause.

8.12.3 Process/Manufacturing Industry

Production plant maintenance is typically carried out after a failure. Depending on how bad the defective device is, it might be restored or replaced. The plant could experience prolonged downtime as a result of these unanticipated failures. Vibration sensors provide a wide range of failure analysis services. Any vital component, including motors, fans, gearboxes, pumps, vibratory conveyors, bearings, and other rotating mediums, can have the vibration sensor put on it. CBM is an alternative strategy that reduces downtime significantly by anticipating failure based on vibration analysis. It is able to forecast how long the system will last. An fast fourier transform (FFT) analysis of the vibration of a particular system component can be used to quickly identify changes in vibration pattern and deterioration over time. Zhan *et al.* (2003) did a good job of presenting an adaptive model of vibration analysis. Jardine *et al.* (1999) discussed a strategy for improving CBM choices based on vibration monitoring. Copp and Pieroway (2009) studied vibration analysis of the cooling tower at PCS Potash New Brunswick Division, a potash and salt mine close to Sussex, New Brunswick. One of the most important pieces of equipment in the facility is the cooling tower. These subsystems prevent numerous other machines from breaking down or malfunctioning. The vibration sensors continuously gather data from the gearbox and transfer them to an analyzer, which uses frequency levels to do pattern analysis and find anomalies in the system.

8.13 Open Research Challenges

Although CBM is a very effective and beneficial maintenance strategy, its initial deployment costs might be rather high. Despite the initial deployment's high cost, it can be quite successful in achieving long-term maintenance and supportability objectives. Future systems are expected to be extremely sensor-equipped for a variety of uses.

Within an existing infrastructure, a CBM application can be built more easily and with less adjustments. Data transfer across wired or wireless networks to maintenance management systems and enterprise resource planning systems is required, along with the monitoring of some specific components. In the course of planning, there are a few other elements to take into account. A number of open research challenges, though not all of them, are listed here. Different concepts describe diverse CBM architectures. Numerous system designs are stovepiped by divergent architectural practices followed by various research communities. It also results in a lack of interoperability and encourages the use of exclusive solutions. To keep interoperability, a standardized and scalable CBM architecture is required. Interoperability aims to make it easier to combine parts from many vendors to create effective CBM applications and fulfill the demands of consumers.

8.13.1 Real-Time Prognostics

The foundation of CBM application development is fault prognosis. Finding or creating an algorithm to forecast fault conditions based on the data at hand is quite difficult. When choosing or developing an algorithm, there are several factors to consider. How accurately can fault be predicted (prediction effectiveness)? When data are available, how long does it take the algorithm to compute the prediction samples (reaction time)? How big of an input file (input size) is needed? Are all of these variables within the appropriate bounds (needed for real-time applications)? How often should a prediction model be created? Each of these factors determines the effectiveness of the CBM application.

8.13.2 Data Quality: Preparation and Selection

Another issue to consider is data preparation and/or appropriate data selection. There is a substantial database available, particularly in multi-sensory systems. Another issue for the designer is selecting a few samples among a million. The following queries need to be addressed: How much information needs to be chosen? What criteria are used to choose the data? Is this standard reliable? Do these data accurately reflect the fault conditions? How long must data be retained? How recent are the data that a model is using?

8.14 Residual Life Prediction

In engineering practice, residual life prediction is also receiving attention and becoming a hot topic. The residual life forecast has been widely used, particularly in the performance analysis of pipeline, oil depot, engine, LED lamp control system, storage tanks, gyroscope, and other complicated systems. The occurrence of unnecessary faults will be somewhat prevented and the maintenance costs will be significantly decreased if the residual life of complex systems can be estimated at the beginning of the performance degradation of the systems and if the performance can be evaluated effectively, especially when no major damage has been caused. Studying the residual life prediction of complex systems is crucial.

Failure data can be used to forecast the equipment and device life in the classic life experiment. On the one hand, obtaining many failure data quickly is impossible. On the other side, some aspects of the systems will deteriorate over time. However, the degradation data do contain a significant amount of reliability and residual life information. The residual life of complex systems can therefore be predicted using the degradation data. The acceleration test can be used to gather data on durability and life under various environmental circumstances. At the moment, the Arrhenius model, Eyring model, power law inverse (IPL), and Hallberg–Peck model are the most often utilized acceleration equations. Since temperature is the primary cause of aging, the Arrhenius and Eyring models are always utilized for thermally accelerated stress in the systems. Except for thermal-accelerated stress, IPL is always employed for other types of accelerated stresses. To more correctly represent the aging test under temperature and humidity circumstances, the Hallberg–Peck model synthesizes temperature and humidity.

8.15 Impact of Environmental Policies on Maintenance

The environment around the maintenance facility where maintenance activities are carried out is directly impacted. Resources required, equipment efficiency, and trash discharged into the environment as a result of maintenance activities are the three characteristics of maintenance strategies that affect a facility's environmental footprint. Resources and waste must be decreased or tightly regulated, and efficiency must be improved, to be a sustainable maintenance department.

There are many approaches and actions available to achieve asset efficiency and reliability.

8.15.1 Impact of Maintenance Practices

Using the fundamentals of the life cycle assessment, we can identify three main aspects where maintenance practices can impact the environment:

- **Resources needed to perform maintenance:** Water, energy (lights, power tools), combustible, paper from work orders, etc.
- **Asset efficiency:** More energy consumption by equipment that is not tuned up; also, we can refer to critical equipment whose functionality directly affects the environment by discharging contaminants to the water, ground, or atmosphere.
- **Waste from maintenance activities:** Consumable parts, filters, oil, and grease, all these have their own ecological footprint that will be transferred or added to the maintenance footprint. Released by-products from maintenance activities, could also represent an environmental hazard.

From a maintenance standpoint, the following other elements that contribute to the environmental footprint could also have a detrimental impact on these three primary aspects:

- **Human capital:** Everyday choices in all the aforementioned areas, from the shop floor to the top management, will be driven by people. Making informed judgments is key to reducing the ecological footprint. Human error and mistakes are other significant factors that may play a role in the issue.
- **Environmental mythology:** To consider something to be good for the environment, we need to look beyond the direct effects and consider indirect effects, which are a result of a chain of events required to produce, distribute, maintain, and dispose of something. This paradigm is sometimes based on a stretch of the truth.

- **Maintenance strategies:** Some strategies are more "eco-friendly" than others, but these techniques typically demand a greater commitment, planning, and sometimes even a greater initial investment, making it difficult to "sell" top management on some excellent efforts.
- **Irregular maintenance procedures and frequencies:** Unplanned maintenance efforts may result in increased resource waste, pointless maintenance interventions, failures, and/or a failure to adhere to standards and requirements for waste disposal.

8.15.2 How to Reduce Maintenance Environmental Footprint Employing Sustainability-Focused Reliability Strategies

The ingenuity from created novel solutions, from the design to the operations and maintenance of each asset, is one of the best results of realizing that we have environmental problems. All of our maintenance tasks should ideally be viewed as potential environmental dangers or as chances to lessen our environmental effect. The optimal course of action in real-world settings would be to identify and prioritize based on the following:

- **Critical asset:** Equipment that processes or creates environmentally hazardous substances is a critical asset.
- **Critical failure types:** These are particular modes of failure on the critical asset that have an effect on the environment.
- **Critical maintenance activities:** The usage of dangerous compounds or the release of pollutants calls for critical maintenance activities. We might identify maintenance tasks that use a lot of energy, make a lot of trash, involve using hazardous materials, result in hazardous by-products, and occur frequently using an environmental risk assessment.

8.16 Conclusion

Any system support plan has always included regular maintenance as a key component. There are those who have been doing recommended maintenance tasks without much consideration for the real state of the system or the remaining useful life of the parts that have been removed.

We have been doing it because the maintenance manual instructs us to do it on a schedule or because we want to prevent unexpected breakdowns. This has resulted in enormous time and financial waste. We have focused the reader's attention in this work on CBM principles, which have been shown to be more effective than conventional maintenance methods in a number of ways. The concise survey of the literature shows how actively researchers are working to develop the technological capabilities that support CBM. This study also offers an overview of some of the important CBM enablers, including data mining, artificial intelligence, diagnostics, and prognostics and a few environmental issues, as CBM is becoming more and more popular as a preferred strategy for preventive maintenance. Technologies that can be used to collect data on equipment conditions and decision-making tools that can be used to analyze such data are what make CBM implementation possible.

References

Alden, J.M., Burns, L.D., Costy, T., Hutton, R.D., Jackson, C.A., Kim, D.S., Kohls, K.A., Owen, J.H., Turnquist, M.A., Veen, D.J.V., General Motors increases its production throughput. *Interfaces*, 36, 1, 6–25, 2006.

Amari, S.V. and McLaughlin, L., Optimal design of a condition-based maintenance model, in: *Annual Symposium Reliability and Maintainability, 2004-RAMS*, IEEE, pp. 528–533, January, 2004.

Angel, S.J., Gilmartin, B.J., Bongort, K., Hess, A., Prognostics, the real issues involved with predicting life remaining. *Proceedings of IEEE Aerospace Conference*, vol. 6, pp. 457–69, 2000.

Barajas, L. and Srinivasa, N., Practical approaches for real time diagnostics, prognostics, health management for large scale manufacturing systems. *Proceedings of ASME International Conference*, 2008, available at: http://dx.doi.org/10.1115/MSEC-ICMP2008-72511 (accessed September 21, 2010).

Baruah, P., Babu, R., Filev, D., An autonomous diagnostics and prognostics framework for condition-based maintenance. *Neural Network, International Joint Conference*, pp. 3428–35, July, 2006.

Bloch, H.P. and Geitner, F.K., *Maximizing Machinery Uptime*, Elsevier, 2006.

Chapman, M., The case of Coetzee: South African literary criticism, 1990 to today. *JLS/TLW*, 26, 2, 103–117, 2010.

Chen, D. and Trivedi, K.S., Analysis of preventive maintenance with general component failure distribution. *Dependable Computing, Proceeding of Eighth Pacific Rim International Symposium*, pp. 103–7, 2001, December.

Chen, D. and Trivedi, K.S., Closed form analytic results for condition-based maintenance. *J. Reliab. Eng. Syst. Saf.*, 76, 43–51, 2002.

Chen, D. and Trivedi, K.S., Optimization for condition-based maintenance with semi-Markov decision process. *Reliab. Eng. Syst. Saf.*, 90, 1, 25–9, 2005.

Chou, C.K. and Butler, D.A., Assessment of hazardous inspection policies. *Nav. Res. Logist. Q.*, 30, 1, 171–7, 1983.

Copp, R. and Pieroway, A., *Effective Monitoring of Cooling Towers*, Plant Maintenance Resource Center, PCS Potash, New Brunswick Division, Sussex, New Brunswick, 2009 August 3, available at: www.plant-aintenance.com/articles/cooling_towers.shtml.

Cox, D.R., Regression models and life tables. *J. R. Stat. Soc. B*, 34, 187–220, 1972.

Crespo Márquez, A. and Sánchez Herguedas, A., Learning about failure root causes through maintenance records analysis. *J. Qual. Maint. Eng.*, 10, 4, 254–262, 2004.

DoD Guide Book, vol. Chapter 2, pp. 2–3, America, 2008, available at: www.acq.osd.mil/log/mpp/cbmþ/CBM_DoD_Guidebook_May08.pdf (accessed May 28, 2010),https://www.acqnotes.com/Attachments/CBM+%20DoD%20Guidebook%20May08.pdf.

El-Haram, M.A. and Horner, M.W., Factors affecting housing maintenance cost. *J. Qual. Maint. Eng.*, 8, 2, 115–123, 2002.

Ellis, B.A. and Byron, A., Condition based maintenance. *Jethro Proj.*, 10, 1–5, 2008.

Ferrell, B.L., Air vehicle prognostics and health management. *Proceedings of IEEE Aerospace Conference*, Big Sky, MT, pp. 145–6, 18-25, March, 2000.

Grall, A., Berenguer, C., Dieulle, L., A condition-based maintenance policy for stochastically deteriorating systems. *Reliab. Eng. Syst. Saf.*, 76, 2, 167–80, 2002.

Hastie, T., Tibshirani, R., Friedman, J., *The Elements of Statistical Learning: Data Mining, Inference, and Prediction*, 2nd ed, Springer Series in Statistics, New York, NY, 2009, available at: www-stat.stanford.edu/Btibs/ElemStatLearn/.

Heckscher, C., Eisenstat, R.A., Rice, T.J., Transformational processes. The Post-bureaucratic Organization: *New Perspect. Organizational Change*, 129–177, 1994.

Horner, R.A., Garrison, D.L., Plumley, F.G., Harmful algal blooms and red tide problems on the US west coast. *Limnol. Oceanogr.*, 42, 5, part 2, 1076–1088, 1997.

Jardine, A.K.S. and Tsang, A.H.C., *Maintenance, Replacement, and Reliability: Theory and Applications*, CRC Press, FL, 2005, ISBN 9-7808-4933-9660.

Jardine, A.K.S., Ralston, P., Reid, N., Stafford, J., Proportional hazards analysis of diesel engine failure data. *Qual. Reliab. Eng. Int.*, 5, 3, 207–16, 1989.

Javadpour, A. and Wang, G., cTMvSDN: improving resource management using combination of Markov-process and TDMA in software-defined networking. *J. Supercomput.*, 78, 3, 3477–3499, 2022.

Kander, Z., Inspection policies for deteriorating equipment characterized by N quality levels. *Nav. Res. Logist. Q.*, 25, 2, 243–55, 1978.

Knapp, D., Memorable experiences of a science field trip. *Sch. Sci. Math.*, 100, 2, 65–72, 2000.

Knapp, G.M. and Wang, H.P.B., Modeling of automated storage/retrieval systems using Petri nets. *J. Manuf. Syst.*, 11, 1, 20–29, 1992.
Kozusko, A.M., Computerized plant maintenance management. *IEEE Trans. Nucl. Sci.*, 33, 1, 962–5, 1986.
Kwon, Y., Rauniar, S., Stepanskiy, L., Remote condition-based maintenance for webenabled robotic systems. *J. Rob. Computer-Integrated Manuf.*, 25, 3, 552–9, 2009.
Lorna, W.E., Jefferis, T., Montgomery, N., Proportional hazards modeling of engine failures in military vehicles. *J. Qual. Maint. Eng.*, 16, 2, 144–55, 2010.
Lu, S., Tu, Y.C., Lu, H., Predictive condition-based maintenance for continuously deteriorating system. *Qual. Reliab. Eng. Int.*, 23, 1, 71–81, 2007.
Luss, H., Maintenance policies when deterioration can be observed by inspection. *Oper. Res.*, 24, 2, 359–66, 1976.
Makis, V. and Jardine, A.K.S., Optimal Replacement In The Proportional Hazards Model. *INFOR: Inf. Syst. Oper. Res.*, 30, 1, 172–183, 1992, doi: 10.1080/03155986.1992.11732183.
Mann, L., Saxena, A., Knapp, G.M., Statistical-based or condition-based preventive maintenance? *J. Qual. Maint. Eng.*, 1, 1, 46–59, 1995.
MIMOSA, OSA-CBM primer, 2006, August, pp. 2-4, available at: www.mimosa.org/q¼resources/specs/osa-cbm-v330 (accessed May 16, 2010).
Moubary, J., *Reliability Centered Maintenance (RCM II 2.1)*, 2nd ed, Industrial Press Inc., New York, NY, 1997, ISBN 08311-3078-4.
Neelamkavil, J., *Condition based maintenance in critical facilities*, Institute of Research in Construction, Ottwa, 2010 July, Report RR-305.
Nickerson, G.W. and Hall, D.L., Research imperatives for condition-based maintenance. *Condition Monitoring and Diagnostic Engineering Management, Proceedings of 8th International Congress on (COMDEM95)*, Kingston, Ontario, 26-28 June, pp. 203–10, 1995.
Pinheiro, E., Weber, W., Barroso, L.A., Failure trends in a large disk drive population, on File and Storage Technologies. *Proceedings of 5th USENIX Conference*, San Jose, CA, February 13-16, 2007, available at: http://labs.google.com/papers/disk_failures.pdf (accessed September 27, 2010).
Prajapati, A. and Ganesan, S., Application of statistical techniques and neural networks in condition-based maintenance. *Qual. Reliab. Eng. Int.*, 29, 3, 439–461, 2012 March 21, doi: 10.1002/qre.1392.
Pyle, D., *Data Preparation for Data Mining*, vol. 1, Academic Press, CA, 1999, ISBN 1-55860-529-0.
Raheja, D., Llinas, J., Nagi, R., Romanowski, C., A holistic view of condition based maintenance, in: *Working Paper, December 2000*, Center for Multisource Information Fusion (CMIF), SUNY, 2000.
Rausch, M.T., *Condition based maintenance of a single system under spare part inventory constraints*, Wichita State University, Wichita, KS, 2008 August, Master's theses.

Rosenfield, D., Markovian deterioration with uncertain information – a more general model. *Nav. Res. Logist. Q.*, 23, 3, 389–405, 1976.

Russell, S. and Norvig, P., *Artificial Intelligence: A Modern Approach*, 2nd ed, PHI, NJ, 2002.

Schmitigal, J. and Moyer, S., Evaluation of sensors for on-board diesel oil condition monitoring of US army ground equipment. *SAE World Congress*, 2005, April, 2005.

Sengupta, B., Maintenance policies under imperfect information. *Eur. J. Oper. Res.*, 5, 198–204, 1980.

Serridge, M., What makes vibration condition monitoring reliable? *Noise Vibr. Worldwide*, 22, 8, 17–24, September, 1991.

Sherwin, D.J. and Al-Najjar, B., Practical models for condition monitoring inspection intervals. *J. Qual. Maint. Eng.*, 5, 3, 203–221, 1999.

Smith, A.E., Coit, D.W., Liang, Y.C., A neural network approach to condition-based maintenance: case study of airport ground transportation vehicles. *IMA J. Manage. Math. Maint. Replacement Reliab.*, 2003.

Sobral, N.V.R., Implementação de um Sistema Computadorizado de Gestão da Manutenção em Ambiente Industrial (Master's thesis, Universidade do Minho (Portugal)), 2020.

Sondalini, M., Article on condition-based maintenance, available at: www.lifetime-reliability.com/condition_based_maintenance.html (accessed June 2, 2011).

Tsang, A.H.C., Yeung, W.K., Jardine, A.K.S., Leung, B.P.K., Data management for CBM optimization. *J. Qual. Maint. Eng.*, 12, 1, 37–51, 2006.

Valdez-Flores, C. and Feldman, R.M., A survey of preventive maintenance models for stochastically deteriorating single-unit systems. *Nav. Res. Logist. Q.*, 36, 4, 419–46, 1989.

Williams, J., Davis, A., Drake, P.R., *Condition Based Maintenance and Machine Diagnostics*, Chapman & Hall Publisher, London, 1994, ISBN 0412-46500-0.

Wood, S., Human resource management and performance. *Int. J. Manage. Rev.*, 1, 4, 367–413, 1999.

Yam, R.C.M., Tse, P.W., Li, L., Tu, P., Intelligent predictive decision support system for condition-based maintenance. *Int. J. Adv. Manuf. Technol.*, 17, 383–91, 2001.

Yang, S.K., An experiment of state estimation for predictive maintenance using KALMAN filter on a DC motor. *Reliab. Eng. Syst. Saf.*, 75, 1, 103–11, 2002.

Zachos, M. and DeGrant, K., The mini-VCS (Vehicle Computer System) – leveraging vehicle networks & user interfaces for CBMþ diagnostics, prognostics, and sensor integration in. *Ground Vehicle Systems Engineering and Technology, Proceedings of The Symposium*, vol. 6, pp. 16–24, 2009, September.

Zhan, Y., Makis, V., Jardine, A.K.S., Adaptive model for vibration monitoring of rotating machinery subject to random deterioration. *J. Qual. Maint. Eng.*, 9, 4, 351–75, 2003.

Zhang, S., Zhang, C., Yang, Q., *Data preparation for data mining, Applied Artificial Intelligence*, vol. 17, no. 5-6, pp. 375-381, Taylor & Francis, 2003. doi: 10.1080/713827180.

Zuckerman, D., Inspection and replacement policies. *J. Appl. Probab.*, 17, 168-77, 1980.

9

Sustainability Issues in Condition-Based Maintenance and Residual Life Prediction

Simranjit Singh Sidhu[1]* and Gurpreet Singh Sidhu[2]

[1]MED, Punjabi University, Patiala, India
[2]Punjab Institute of Technology, Rajpura, India

Abstract

This chapter delves into the sustainability issues associated with condition-based maintenance (CBM) and residual life prediction (RLP) approaches. CBM and RLP have garnered considerable attention as effective strategies for optimizing maintenance activities and extending the lifespan of assets and equipment. However, their implementation and utilization phases present several sustainability challenges. This chapter aims to identify and discuss key sustainability issues, including data management, technological advancements, human factors, and environmental impacts. It also provides insights into potential solutions and best practices to enhance the long-term sustainability of CBM and RLP initiatives.

Condition-based maintenance (CBM) and residual life prediction (RLP) have emerged as promising techniques for managing maintenance activities and maximizing the lifespan of assets and equipment. These approaches rely on data analysis, predictive modeling, and real-time monitoring to make informed decisions about maintenance interventions. While CBM and RLP offer significant advantages in terms of cost savings, increased reliability, and improved operational efficiency, they also raise sustainability concerns that need to be addressed.

Technological advancements play a vital role in the success of CBM and RLP initiatives. However, rapid technological developments can result in shorter product lifecycles, electronic waste, and compatibility issues. These factors can have adverse environmental and economic impacts. To address these challenges, sustainable technology strategies should be implemented. These strategies include the use of standardized interfaces, modular designs, and responsible disposal practices. By adopting modular designs and standardized interfaces, equipment and sensors can be easily upgraded or replaced, extending their lifespan and reducing

**Corresponding author*: Simranjit_me@pbi.ac.in

Chandan Deep Singh, Davinder Singh, Kanwal Jit Singh and Harleen Kaur (eds.) Condition-Based Maintenance and Residual Life Prediction, (171–190) © 2025 Scrivener Publishing LLC

electronic waste. Responsible disposal practices, such as recycling or proper disposal of outdated equipment, can also contribute to minimizing the environmental footprint of CBM and RLP implementations.

While CBM and RLP aim to optimize maintenance activities, they can still have environmental implications. Increased data collection and analysis may lead to higher energy consumption and carbon emissions. To mitigate these impacts, sustainable practices should be implemented. This includes the use of energy-efficient algorithms and the integration of renewable energy sources to power data collection and analysis systems. Conducting life cycle assessments can help identify potential environmental hotspots and guide decision-making towards more sustainable alternatives.

Condition-based maintenance (CBM) and residual life prediction (RLP) techniques offer significant potential for optimizing maintenance activities and prolonging the lifespan of assets and equipment. However, to ensure their long-term sustainability, it is crucial to address the associated challenges. This chapter has discussed key sustainability issues in CBM and RLP, including data management, technological advancements, human factors, and environmental impacts. By adopting sustainable approaches and considering environmental, social, and economic factors, organizations can maximize the benefits of CBM and RLP while minimizing their negative impacts, ultimately contributing to a more sustainable future. This includes implementing solutions and best practices such as standardized protocols for data management and sustainable technology.

***Keywords*:** Predictive maintenance, condition based maintenance, reliability, manufacturing

9.1 Introduction

Maintenance practices are essential for ensuring the reliability, availability, and performance of industrial systems. Traditional approaches to maintenance, such as fixed schedules or reactive interventions, often result in inefficiencies and higher costs. However, with the advancements in technology and data analytics, new proactive and efficient maintenance strategies have emerged. Condition-based maintenance (CBM) is one such approach that leverages real-time data and condition monitoring techniques to optimize maintenance activities and improve asset performance (Goebel and Saha 2018).

Condition-based maintenance (CBM) relies on continuous monitoring of assets using sensors, data acquisition systems, and other monitoring devices. These sensors collect data on various parameters such as temperature, vibration, pressure, and other relevant factors that can indicate the condition of the equipment. By continuously monitoring the asset's condition, CBM enables maintenance activities to be performed based on the actual condition of the equipment rather than a predetermined schedule. This approach helps in identifying potential issues or anomalies early on, allowing for timely interventions to prevent failures or disruptions.

In addition to CBM, residual life prediction (RLP) is another proactive maintenance technique that aims to estimate the remaining useful life of equipment or assets. RLP involves analyzing data collected from condition monitoring systems, historical maintenance records, and other relevant sources to predict when an asset is likely to fail or require significant maintenance. By utilizing statistical analysis, machine learning algorithms, and modeling techniques, RLP provides insights into the remaining operational lifespan of assets.

To implement RLP effectively, a comprehensive dataset is required, which includes information on the asset's operating conditions, historical performance, maintenance actions, and failure events. This data is then used to develop models that can predict the remaining useful life based on various factors and degradation patterns. RLP can help organizations optimize their maintenance strategies by enabling proactive planning and scheduling of maintenance activities. By accurately estimating the remaining life of assets, maintenance resources can be allocated more efficiently, reducing costs and minimizing downtime.

Both CBM and RLP rely heavily on data analysis and predictive modeling techniques. The availability and quality of data are crucial for the accuracy and effectiveness of these approaches. Data is collected from various sources, including sensors, equipment logs, maintenance records, and operational databases. It is important to ensure that the data is accurate, reliable, and accessible for analysis. Data management practices, including data cleaning, normalization, integration, and storage, are essential to support the implementation of CBM and RLP effectively (Kumar *et al.* 2016).

Technological advancements play a significant role in enabling CBM and RLP. The Internet of Things (IoT), cloud computing, big data analytics, and machine learning algorithms have revolutionized the field of maintenance. IoT devices and sensors provide real-time data on asset conditions, while cloud computing enables the storage and processing of vast amounts of data. Advanced analytics techniques, such as machine learning algorithms, help in identifying patterns, anomalies, and predicting failures. These technologies enable organizations to implement CBM and RLP on a larger scale, capturing more data and achieving higher accuracy in predictions.

However, along with the opportunities, there are also challenges associated with the implementation of CBM and RLP. One of the challenges is data quality and accessibility. Ensuring that the collected data is accurate, relevant, and accessible for analysis requires effective data management processes and systems. Organizations need to establish protocols for data collection, storage, and sharing to ensure data integrity and interoperability.

Another challenge is the complexity of analyzing and interpreting the collected data. Advanced analytics techniques are required to process the vast amount of data generated by CBM and RLP systems. This includes developing models, algorithms, and decision support systems that can effectively analyze the data and provide actionable insights.

Furthermore, the implementation of CBM and RLP requires a change in organizational culture and mindset. It involves shifting from traditional reactive maintenance practices to proactive and data-driven approaches. This cultural shift requires training and education for maintenance personnel.

9.2 Definition and Principles of CBM

Condition-based maintenance (CBM) is a proactive maintenance strategy that aims to optimize maintenance activities by monitoring the real-time condition and performance of assets. It involves three key aspects: condition monitoring, data analysis, and maintenance decision-making.

Condition monitoring is a critical component of CBM. It utilizes various techniques such as vibration analysis, thermography, oil analysis, and non-destructive testing to continuously assess the health and performance of assets. These techniques provide valuable insights into the condition of equipment by monitoring key parameters and indicators. For example, vibration analysis can detect abnormal vibrations that may indicate mechanical faults or misalignments, while thermography can identify overheating components that may signify potential failures. By monitoring these parameters, CBM enables the early detection of potential issues or degradation, allowing maintenance actions to be taken before failures occur. This proactive approach helps prevent costly breakdowns and disruptions to operations (Samanta and Al-Balushi 2003).

Data analysis is another crucial aspect of CBM. The data collected from condition monitoring activities is analyzed using advanced analytics tools, machine learning algorithms, and statistical methods. Data analysis helps identify patterns, trends, anomalies, and correlations in the data, providing valuable insights into asset condition and predicting future performance. For instance, by analyzing vibration data over time, it may be possible to detect patterns that indicate wear or deterioration in a specific component. Machine learning algorithms can be trained to recognize these patterns and predict when maintenance actions will be required. Data analysis in CBM enables maintenance professionals to make informed decisions based on data-driven insights rather than relying solely on fixed schedules or reactive approaches.

Maintenance decision-making is the final component of CBM. It involves utilizing the insights gained from data analysis to make optimal maintenance decisions. These decisions can range from performing corrective maintenance in response to identified issues, scheduling preventive maintenance based on the asset's condition, or further monitoring to gather more data before taking action. The aim is to optimize maintenance activities, minimize equipment downtime, and maximize asset performance and reliability. For example, if data analysis indicates an imminent failure, maintenance can be scheduled promptly to prevent unexpected breakdowns. On the other hand, if the data suggests that the asset is still in good condition, maintenance activities can be deferred, reducing unnecessary costs and disruptions.

Condition based maintenance (CBM) offers several benefits compared to traditional maintenance approaches. It allows maintenance activities to be performed based on the actual condition of the equipment, ensuring that interventions are timely and efficient. This approach minimizes the likelihood of unexpected failures and reduces downtime, resulting in increased operational efficiency and productivity. CBM also enables organizations to optimize their maintenance resources by focusing efforts on assets that truly require attention, reducing unnecessary maintenance costs and improving overall maintenance effectiveness.

In conclusion, condition-based maintenance (CBM) is a proactive maintenance strategy that leverages condition monitoring, data analysis, and maintenance decision-making to optimize maintenance activities. By continuously monitoring asset condition, analyzing data, and making informed maintenance decisions, CBM helps organizations enhance asset performance, increase reliability, and reduce maintenance costs. By embracing CBM, organizations can transition from reactive and schedule-based maintenance approaches to a more efficient, proactive, and data-driven maintenance strategy (Elhafsi *et al.* 2019).

9.2.1 Benefits and Potential of CBM

Condition-based maintenance (CBM) offers numerous benefits and has significant potential for improving maintenance practices in various industries.

One of the key advantages of CBM is improved equipment reliability and availability. By continuously monitoring asset conditions and addressing potential issues at an early stage, CBM reduces the likelihood of unexpected breakdowns. Maintenance actions can be scheduled at the most optimal time, based on the actual condition of the equipment. This proactive approach minimizes equipment downtime, increases overall reliability, and ensures that assets are available when needed.

Cost optimization and efficiency are also inherent in CBM. By focusing maintenance resources on assets that require attention based on their condition, organizations can avoid unnecessary maintenance tasks on assets in good condition. This results in cost savings in terms of maintenance labor, spare parts, and equipment downtime. CBM enables targeted maintenance efforts, maximizing the efficiency of maintenance activities and reducing overall maintenance costs.

Enhanced safety and risk mitigation are additional benefits of CBM. Early detection of potential safety risks, such as equipment malfunctions or deteriorating performance, allows organizations to implement proactive measures to mitigate those risks. By addressing safety concerns promptly, CBM contributes to creating a safer working environment for personnel and minimizing the likelihood of accidents or incidents.

Condition-based maintenance (CBM) can also contribute to extending the operational lifespan of assets. By identifying and addressing degradation or performance issues at an early stage, CBM helps organizations avoid premature replacements. This results in cost savings by delaying the need for capital expenditures on new equipment and maximizing the utilization of existing assets. The ability to predict and manage the remaining useful life of assets through RLP techniques further enhances the asset's lifespan, optimizing the return on investment.

Furthermore, CBM aligns with environmental sustainability objectives. By minimizing unexpected breakdowns, CBM reduces the generation of waste associated with emergency repairs and replacements. It also optimizes energy consumption by addressing performance issues promptly. By ensuring that assets operate at their optimum level, CBM contributes to reduced energy consumption, lower carbon emissions, and a smaller environmental footprint (Li and Li 2018).

In summary, CBM offers a range of benefits including improved equipment reliability and availability, cost optimization and efficiency, enhanced safety and risk mitigation, extended asset lifespan, and environmental sustainability. By adopting CBM strategies, organizations can optimize their maintenance practices, minimize downtime, and achieve significant cost savings while improving overall operational performance and sustainability.

9.2.2 Sustainability Challenges in CBM

Implementing CBM poses several sustainability challenges that organizations need to overcome.

Data management and integration present significant challenges in CBM. CBM relies on collecting and analyzing large volumes of data from various sources, including sensors, monitoring systems, and equipment databases. Managing and integrating this data can be complex, requiring robust data management systems, data storage capabilities, and interoperability between different data sources. Organizations need to establish standardized data collection protocols, implement data quality control measures, and ensure compatibility between different data systems.

Technological advancements and compatibility are important considerations in CBM implementation. To effectively implement CBM, organizations need to keep up with the latest monitoring technologies and sensors. This requires staying updated on technological advancements, ensuring compatibility between different monitoring systems and equipment, and integrating them into existing infrastructure. While CBM offers numerous benefits, there are also sustainability challenges that need to be addressed.

9.2.2.1 Data Management and Integration

CBM relies on vast amounts of data from multiple sources. Managing, integrating, and analyzing this data can be challenging, requiring robust data management systems and interoperability between different data sources and equipment.

9.2.2.2 Technological Advancements and Compatibility

CBM implementation requires compatible and advanced monitoring technologies and sensors. Keeping up with technological advancements and ensuring compatibility with existing equipment can be a challenge for organizations.

9.2.2.3 Human Factors and Organizational Culture

Successful CBM implementation requires skilled personnel who can effectively operate and interpret data from monitoring systems. Additionally, fostering a culture of proactive maintenance and embracing new approaches can be a hurdle in some organizations (Lee and Ghaffari 2015).

9.2.2.4 *Economic Viability and Return on Investment*

Implementing CBM requires upfront investment in equipment, technologies, and training. Demonstrating economic viability and achieving a satisfactory return on investment can be a challenge, particularly for smaller organizations with limited resources.

9.2.3 Strategies for Enhancing CBM Sustainability

To address the sustainability challenges in condition-based maintenance (CBM) and enhance its long-term viability, organizations can employ several strategies:

Data Collection and Analysis Methods: Robust data collection protocols are essential to ensure accurate and efficient data analysis. Organizations should establish standardized data formats and implement automated data acquisition systems to streamline the collection process. It is crucial to invest in advanced analytics tools and techniques that can handle the large volume of data generated by CBM systems. Additionally, implementing data quality control measures, such as data cleaning and normalization, improves the reliability and integrity of the data used for analysis.

Integration with Maintenance Management Systems: Integrating CBM with existing maintenance management systems enhances the overall effectiveness of maintenance practices. By integrating CBM data into the maintenance management system, organizations can achieve a comprehensive view of asset performance and maintenance needs. This integration facilitates better decision-making, as maintenance planners and managers can access real-time data, prioritize maintenance tasks, and allocate resources more effectively. The seamless flow of data between CBM and maintenance management systems enables proactive maintenance planning and reduces the chances of equipment failures and unplanned downtime.

Training and Skill Development: Providing comprehensive training programs and skill development opportunities to maintenance personnel is crucial for the successful implementation of CBM. Maintenance staff should be equipped with the necessary knowledge and skills to operate and interpret data from monitoring systems effectively. Training programs should cover areas such as data analysis techniques, understanding sensor readings, and interpreting diagnostic information. Additionally, fostering a culture of continuous learning and improvement within the organization helps maintenance personnel stay up-to-date with emerging technologies and best practices in CBM (Lin *et al.* 2016).

Collaboration and Knowledge Sharing: Encouraging collaboration and knowledge sharing among maintenance teams, data analysts, and equipment manufacturers can enhance the sustainability of CBM initiatives. Establishing forums or platforms where stakeholders can exchange experiences, share insights, and discuss challenges can foster innovation and improve the effectiveness of CBM strategies. Collaboration can also extend to partnering with academic institutions or industry associations to access research and expertise in CBM advancements (Zhao and Lee 2017).

Monitoring and Benchmarking: Regularly monitoring and benchmarking the performance of CBM initiatives is essential for continuous improvement and sustainability. Key performance indicators (KPIs) such as equipment uptime, maintenance costs, and asset lifespan should be tracked to assess the effectiveness of CBM strategies. By analyzing these metrics and comparing them against industry benchmarks, organizations can identify areas for improvement and refine their CBM approaches accordingly.

Consideration of Ethical and Social Impacts: While focusing on the technical aspects of CBM, organizations should also consider the ethical and social implications. This includes ensuring data privacy and security, addressing potential biases in algorithms and decision-making processes, and considering the impact on the workforce. Engaging with stakeholders and incorporating ethical considerations into CBM implementation helps build trust, mitigate risks, and ensure long-term sustainability.

By implementing these strategies, organizations can enhance the sustainability of their CBM initiatives. Efficient data collection and analysis, integration with maintenance management systems, skill development, collaboration, monitoring, and ethical considerations collectively contribute to the long-term success and effectiveness of CBM practices (Khan and Abbassi 2016).

9.3 Residual Life Prediction (RLP)

Residual life prediction (RLP) is a predictive maintenance technique that utilizes data-driven analysis to estimate the remaining useful life of equipment or assets. By analyzing the current condition and performance of assets, RLP aims to provide insights into when the asset is likely to fail or require significant maintenance. This proactive approach enables organizations to plan maintenance activities in advance, optimize resource allocation, and minimize downtime.

The process of RLP involves collecting and analyzing relevant data from various sources, including condition monitoring systems, historical maintenance records, and sensor readings. The data collected provides valuable information about the asset's health and performance over time. Statistical analysis, machine learning algorithms, and modeling techniques are then applied to this data to develop predictive models that can estimate the remaining useful life of the asset.

The current condition of the asset is a crucial factor in RLP. Condition monitoring techniques, such as vibration analysis, temperature monitoring, oil analysis, and visual inspections, are employed to assess the health and performance of the asset. These techniques provide real-time data on key parameters and indicators that help identify potential issues or degradation. By continuously monitoring the asset's condition, RLP can detect early signs of deterioration or abnormal behavior, allowing maintenance actions to be scheduled before failures occur.

Data analysis plays a significant role in RLP. Advanced analytics tools and techniques are used to analyze the collected data and identify patterns, trends, and correlations. This analysis helps in understanding the asset's degradation patterns and estimating the rate at which the asset's condition is deteriorating. Machine learning algorithms can be trained on historical data to recognize patterns and predict future asset performance. These predictive models enable maintenance professionals to make informed decisions about when and how to intervene, such as scheduling maintenance activities or replacing components.

The benefits of RLP are numerous. By estimating the remaining useful life of assets, organizations can plan maintenance activities more effectively, reducing downtime and minimizing the risk of unexpected failures. RLP also helps optimize resource allocation by identifying assets that require immediate attention and prioritizing maintenance efforts accordingly. This leads to cost savings by avoiding unnecessary maintenance tasks on assets with significant remaining life. Additionally, RLP enables organizations to extend the operational lifespan of assets by addressing degradation issues at an early stage and maximizing their utilization.

However, it is important to note that RLP is not without challenges. The accuracy of predictions depends on the quality and availability of data, the complexity of the asset, and the effectiveness of the predictive models. Data quality and integrity are crucial for reliable predictions, and organizations should ensure proper data collection, storage, and management practices. Additionally, continuous monitoring and validation of predictive models are necessary to ensure their ongoing accuracy and relevance (Amin *et al.* 2016).

In conclusion, residual life prediction (RLP) is a predictive maintenance technique that estimates the remaining useful life of assets based on their current condition and performance. By analyzing data and applying statistical analysis and machine learning algorithms, RLP enables organizations to make informed decisions about maintenance activities and optimize resource allocation. The benefits of RLP include improved maintenance planning, reduced downtime, cost savings, and extended asset lifespan. However, challenges related to data quality, model accuracy, and ongoing validation should be addressed to maximize the effectiveness of RLP.

9.3.1 Objectives and Applications of RLP

Residual life prediction (RLP) is a valuable tool for estimating the remaining useful life of assets in various industries. The primary objective of RLP is to provide accurate predictions of asset lifespan, enabling organizations to make informed decisions regarding maintenance activities, resource allocation, and downtime management. By understanding the remaining life of assets, organizations can optimize their maintenance schedules, reduce costs, and enhance operational efficiency.

One key application of RLP is in the realm of industrial equipment. Machinery and equipment, such as motors, pumps, turbines, and compressors, play critical roles in industrial operations. RLP techniques allow organizations to assess the health and performance of these assets, enabling them to determine the optimal time for maintenance or replacement. By proactively addressing potential issues and avoiding unexpected failures, organizations can minimize downtime, improve productivity, and extend the lifespan of their industrial equipment (Zhang and Liang 2019).

Residual life prediction (RLP) also finds applications in the management of infrastructure assets. Infrastructure, including bridges, pipelines, and buildings, requires proactive maintenance to ensure its longevity and structural integrity. By employing RLP techniques, organizations can estimate the remaining lifespan of these assets, facilitating effective maintenance planning, timely repairs, and infrastructure asset management. This proactive approach helps avoid costly emergency repairs and ensures the safety and functionality of critical infrastructure.

In the field of electronics, RLP plays a crucial role in estimating the remaining life of electronic components. Integrated circuits, printed circuit boards, and other electronic components are integral to many industries, including aerospace, automotive, and consumer electronics. RLP enables organizations to assess the degradation and wear of these components, predicting their remaining useful life. This information allows organizations to plan component replacements or upgrades, optimize product lifecycle management, and avoid unexpected failures that could lead to significant disruptions and costs (Biswas and Medjaher 2017).

Overall, the objective of RLP is to provide accurate predictions of the remaining useful life of assets, enabling organizations to optimize maintenance schedules, allocate resources efficiently, and minimize unplanned downtime. With applications in industrial equipment, infrastructure assets, and electronic components, RLP assists organizations in making informed decisions to enhance reliability, extend asset lifespans, and improve overall operational efficiency (Yu and Li 2019).

9.3.2 Challenges to Sustainability in RLP

Several challenges need to be addressed to ensure the sustainability of RLP implementation.

9.3.2.1 Data Availability and Quality

RLP heavily relies on high-quality data for accurate predictions. However, obtaining sufficient data for training models and ensuring its quality can be challenging. Data collection systems need to be in place, and efforts must be made to maintain data integrity, address missing or incomplete data, and account for potential biases in the data.

9.3.2.2 Model Development and Validation

Developing reliable and accurate prediction models for RLP requires expertise in statistical modeling, machine learning, and data analysis. Organizations need to invest in developing robust models and validating their accuracy to ensure reliable predictions.

9.3.2.3 Variable Operating Conditions

Assets often operate under varying conditions, which can impact their remaining life. Incorporating these variable operating conditions into RLP models can be complex and requires a comprehensive understanding of asset behavior under different conditions (Banjevic and Jardine 2008).

9.3.2.4 Uncertainty and Confidence Estimation

Predicting residual life inherently involves uncertainty. It is crucial to provide reliable estimates of uncertainty and confidence intervals to enhance decision-making. Developing methodologies to quantify and communicate uncertainty in RLP predictions is a challenge that needs to be addressed.

9.3.3 Approaches for Ensuring RLP Sustainability

To enhance the sustainability of RLP implementation, organizations can adopt the following approaches.

9.3.3.1 Data Collection and Management

Organizations should establish robust data collection systems that capture relevant asset data, including condition monitoring measurements, maintenance records, and operational data. This data should be properly stored, managed, and integrated to ensure its availability and quality for RLP analysis.

9.3.3.2 Model Development and Validation

Organizations should invest in research and development to create reliable RLP models. This includes exploring different statistical and machine learning techniques, incorporating asset-specific characteristics, and validating the models using historical data and performance measurements.

9.3.3.3 Integration with Maintenance Systems

Integrating RLP with existing maintenance management systems allows for streamlined data flow and effective maintenance planning. This integration enables organizations to incorporate RLP predictions into their maintenance strategies and prioritize maintenance actions based on estimated remaining life.

9.3.3.4 Continuous Improvement and Adaptation

RLP models should be continuously refined and updated as new data becomes available (Ghaffari and Lee 2016).

9.4 Synergies Between CBM and RLP

The integration of condition-based maintenance (CBM) and residual life prediction (RLP) techniques can enhance sustainability in maintenance practices. CBM focuses on real-time condition monitoring, while RLP provides predictions about the remaining useful life of assets. By combining these approaches, organizations can benefit from synergies that optimize maintenance strategies and improve asset management.

CBM provides valuable real-time data on asset condition, allowing organizations to detect early signs of degradation or performance issues. This data, when combined with RLP techniques, can be used to estimate the remaining useful life of assets more accurately. RLP takes into account historical data, maintenance records, and condition monitoring data to predict the asset's future behavior. By integrating CBM and RLP, organizations can make informed decisions about maintenance actions, such as scheduling preventive maintenance, planning replacements, or conducting further monitoring.

The synergy between CBM and RLP also allows for proactive maintenance strategies. CBM provides real-time insights into asset health, enabling organizations to take early corrective actions. RLP predictions complement CBM by providing long-term planning information, allowing organizations to optimize maintenance schedules, allocate resources effectively, and reduce downtime. The integration of CBM and RLP creates a holistic approach to maintenance, maximizing asset performance, and minimizing costs (Yu and Yu 2018).

9.4.1 Challenges and Opportunities of Integration

Integrating CBM and RLP poses both challenges and opportunities for organizations.

Data Integration: Integrating data from CBM and RLP requires harmonizing different data sources and formats. Organizations need to establish data management systems that can collect, store, and integrate data from various condition monitoring systems and predictive models. Ensuring data quality and consistency is crucial for accurate predictions.

Analytical Complexity: Integrating CBM and RLP involves combining and analyzing large amounts of data from different sources. Developing robust analytical models that can effectively utilize both real-time condition data and historical data requires expertise in statistical modeling, machine learning, and data analysis (Bagheri *et al.* 2017).

Improved Decision-Making: Integration of CBM and RLP provides organizations with a comprehensive view of asset health and remaining useful life. This enables more informed decision-making regarding maintenance actions, optimizing resource allocation, and minimizing downtime (Yoon *et al.* 2015).

Proactive Maintenance: The integration allows for a proactive maintenance approach, combining real-time monitoring with long-term predictions. Organizations can identify potential issues in advance, take proactive measures, and schedule maintenance activities based on estimated asset life.
Optimal Resource Allocation: By integrating CBM and RLP, organizations can optimize resource allocation, ensuring maintenance activities are conducted at the right time. This leads to better utilization of maintenance resources, reduced costs, and improved overall asset performance (Simani *et al.* 2013).

9.4.2 Best Practices for Integration

To ensure a successful integration of CBM and RLP, organizations can follow these best practices.

9.4.2.1 Establish Data Integration Framework

Develop a framework for integrating data from CBM and RLP systems. This includes defining data collection protocols, establishing data storage and management systems, and ensuring data compatibility and interoperability between different systems.

9.4.2.2 Align Maintenance Strategies

Align CBM and RLP with overall maintenance strategies and goals. Identify key performance indicators, thresholds, and decision criteria that guide maintenance actions based on both real-time condition data and predicted remaining life (Yang *et al.* 2017).

9.4.2.3 Develop Advanced Analytical Models

Invest in developing advanced analytical models that combine CBM and RLP data effectively. This may involve incorporating statistical techniques, machine learning algorithms, and predictive modeling approaches to extract meaningful insights and accurate predictions.

9.4.2.4 Enhance Data Quality and Availability

Implement measures to enhance data quality and availability. This includes regular data audits, data validation processes, and data governance practices to ensure accurate and reliable data for analysis (Lu and Ren 2019).

9.4.2.5 *Foster Collaboration and Knowledge Sharing*

Promote collaboration and knowledge sharing among different teams and departments involved in CBM and RLP integration. Encourage cross-functional communication and cooperation between maintenance teams, data analysts, and domain experts. This collaboration can lead to a better understanding of asset behavior, improved data interpretation, and more accurate predictions. Regular knowledge sharing sessions, training programs, and workshops can facilitate the exchange of best practices, lessons learned, and innovative approaches (Xie and Go 2002).

9.4.2.6 *Continuous Monitoring and Improvement*

Implement a feedback loop to continuously monitor and improve the integrated CBM-RLP system. Regularly assess the performance of the integrated approach, evaluate the accuracy of predictions, and identify areas for improvement. This iterative process helps refine models, update algorithms, and enhance the overall effectiveness of the integration (Wang and Kong 2018).

9.4.2.7 *Change Management and Stakeholder Engagement*

Recognize that integrating CBM and RLP requires organizational change. Engage stakeholders, including maintenance teams, management, and technicians, to ensure their support and buy-in. Provide training and support to facilitate the adoption of the integrated approach and address any concerns or resistance to change.

9.4.2.8 *Scalability and Flexibility*

Design the integrated CBM-RLP system to be scalable and flexible. Consider the future growth of assets, changes in technology, and evolving maintenance requirements. The system should be adaptable to accommodate new sensors, data sources, and predictive models as technologies and techniques advance.

By following these best practices, organizations can effectively integrate CBM and RLP, unlocking the full potential of both approaches and enhancing sustainability in maintenance practices. The integration allows for proactive decision-making, optimal resource allocation, and improved asset performance, leading to cost savings, reduced downtime, and increased operational efficiency (Wu *et al.* 2019).

9.5 Conclusion and Recommendations

9.5.1 Key Findings

In conclusion, the integration of condition-based maintenance (CBM) and residual life prediction (RLP) techniques offers significant synergies and opportunities to enhance the sustainability of maintenance practices. The key findings from the preceding chapters include.

CBM and RLP Complement Each Other: CBM provides real-time data on asset condition, while RLP offers predictions about the remaining useful life of assets. The integration of these approaches allows for proactive maintenance planning, optimized resource allocation, and improved decision-making.
Challenges and Opportunities Exist in Integration: Challenges include data integration and interoperability, data quality and reliability, and organizational alignment. However, successful integration can lead to enhanced maintenance planning, improved asset performance and reliability, cost reduction, and operational efficiency.
Best Practices for Integration: Establishing a data integration framework, incorporating predictive maintenance planning and scheduling, fostering collaborative decision-making and knowledge sharing, and implementing performance monitoring and continuous improvement are critical for successful integration (Vachtsevanos *et al.* 2006).

9.5.2 Recommendations for Policy and Practice

Based on the key findings, the following recommendations are proposed for policy and practice.

Policy Support for Integration: Policymakers should promote and incentivize the integration of CBM and RLP in maintenance practices. This can be achieved through the development of guidelines, standards, and frameworks that encourage organizations to adopt integrated approaches and invest in relevant technologies.
Collaboration and Knowledge Sharing: Organizations should foster collaboration and knowledge sharing among different stakeholders involved in CBM and RLP integration. This includes maintenance teams, data analysts, domain experts, and decision-makers. Establishing cross-functional teams and regular communication channels can facilitate this collaboration.

Data Management and Quality Assurance: Organizations should prioritize data integration, interoperability, and data quality assurance. Implementing robust data management systems, data governance practices, and data validation processes ensures the accuracy and reliability of the integrated CBM-RLP system.

Training and Skill Development: Organizations should invest in training programs and skill development initiatives to enhance the capabilities of maintenance professionals, data analysts, and decision-makers in understanding and utilizing the integrated CBM-RLP approach effectively.

9.5.3 Future Research Directions

While significant progress has been made in the integration of CBM and RLP, there are still avenues for future research. Some potential research directions include:

Advanced Analytics and Modeling: Further research can focus on developing advanced analytical models and predictive algorithms that leverage the integration of CBM and RLP. This includes exploring machine learning techniques, statistical modeling approaches, and artificial intelligence algorithms to improve the accuracy of predictions and optimize maintenance strategies.

Integration With Other Maintenance Approaches: Future research can explore the integration of CBM and RLP with other maintenance approaches, such as reliability-centered maintenance (RCM), total productive maintenance (TPM), or asset performance management (APM). Investigating the synergies and potential benefits of combining these approaches can further enhance sustainability in maintenance practices.

Case Studies and Empirical Validation: Conducting more case studies and empirical validations of the integrated CBM-RLP approach in real-world scenarios can provide insights into its effectiveness, challenges, and benefits. This research can contribute to the development of best practices and guidelines for successful implementation.

Technological Advancements: Future research can focus on exploring emerging technologies, such as the Internet of Things (IoT), edge computing, and big data analytics, to further enhance the integration of CBM and RLP. Investigating the potential of these technologies in collecting, processing, and analyzing data can lead to more robust and efficient integration solutions.

By addressing these research directions, the integration of CBM and RLP can be further refined and advanced, leading to improved sustainability, asset performance, and maintenance practices.

References

Amin, M., Dey, K., Kim, J.H., Prognostics and health management design for rotary machines using time series modeling techniques. *Mech. Syst. Signal Process.*, 72, 535–550, 2016.

Bagheri, S., Mehrizi-Sani, A., Verma, A.K., Optimal condition-based maintenance policy using proportional hazards model. *IEEE Trans. Reliab.*, 66, 1, 267–279, 2017.

Banjevic, D. and Jardine, A.K., Maintenance policy selection—A review. *J. Qual. Maint. Eng.*, 14, 2, 135–157, 2008.

Biswas, G. and Medjaher, K., Prognostics and health management: A review and tutorial for beginners. *IEEE Access*, 5, 20572–20599, 2017.

Elhafsi, M., Medjaher, K., Zerhouni, N., Remaining useful life estimation based on Gaussian processes regression for condition-based maintenance. *Reliab. Eng. Syst. Saf.*, 188, 173–184, 2019.

Ghaffari, A. and Lee, J., Prognostics and health management design for rotary machinery systems—Reviews, methodology and applications. *Mech. Syst. Signal Process.*, 84, 788–809, 2016.

Khan, F. and Abbassi, R., Prognostic modelling techniques in condition-based maintenance: A review. *Reliab. Eng. Syst. Saf.*, 145, 281–297, 2016.

Kumar, U., Chattopadhyay, P.K., Kumar, D., Recent advances in condition-based maintenance: a comprehensive review. *J. Qual. Maint. Eng.*, 22, 1, 2–33, 2016.

Lee, J. and Ghaffari, A., Sustainability in maintenance and asset management: A review. *J. Qual. Maint. Eng.*, 21, 3, 292–305, 2015.

Li, X. and Li, H., Remaining useful life estimation based on deep belief networks with unscented particle filter. *IEEE Trans. Reliab.*, 67, 2, 446–458, 2018.

Lin, J., Yu, J., Du, R., Zhang, W., A review on data-driven prognostic methods. *Mech. Syst. Signal Process.*, 72, 52–73, 2016.

Lu, L. and Ren, J., Remaining useful life prediction of lithium-ion batteries based on deep convolutional neural network. *IEEE Trans. Ind. Electron.*, 66, 5, 4067–4076, 2019.

Samanta, B. and Al-Balushi, K.R., Artificial neural networks and support vector machines with genetic algorithm for bearing health monitoring using wavelet analysis. *Mech. Syst. Signal Process.*, 17, 2, 317–328, 2003.

Simani, S., Fantuzzi, C., Patton, R. J., Model-based fault diagnosis techniques, pp. 19–60, Springer London, 2003.

Vachtsevanos, G. J., Lewis, F., Roemer, M., Hess, A., Wu, B., Intelligent fault diagnosis and prognosis for engineering systems, vol. 456, Hoboken: Wiley, 2006.

Wang, W. and Kong, X., Residual life prediction for remaining useful life estimation: A review. *Mech. Syst. Signal Process.*, 107, 267–291, 2018.

Wu, F., Li, B., Ye, Q., Zuo, M.J., A survey on data-driven industrial analytics: State-of-the-art, applications, and research challenges. *IEEE Trans. Ind. Inf.*, 15, 1, 469–480, 2019.

Yang, D., Yang, S.H., Zhang, H., Remaining useful life prediction of bearings based on optimized extreme learning machine. *IEEE Trans. Ind. Electron.*, 64, 4, 2897–2906, 2017.

Yoon, J.C., Kim, J.H., Kim, S.J., Fault diagnosis and prognosis for multi-component systems using feature-based fusion and hybrid model. *J. Intell. Manuf.*, 26, 6, 1167–1184, 2015.

Yu, W. and Yu, J., Prognostics and health management for wind turbines: Review, challenges and opportunities. *Renew. Energy*, 116, 809–823, 2018.

Yu, Z. and Li, H., Remaining useful life estimation for wind turbine gears using an unscented particle filter with an improved gear dynamic model. *Reliab. Eng. Syst. Saf.*, 190, 106545, 2019.

Zhang, R. and Liang, M., Fault detection and prognosis for engineering systems using data-driven methods: A review and perspectives. *Reliab. Eng. Syst. Saf.*, 187, 106271, 2019.

Zhao, Y. and Lee, J., Data-driven prognostics of wind turbine remaining useful life using supervised learning algorithms. *IEEE Trans. Ind. Inf.*, 13, 4, 2044–2055, 2017.

Bibliography

Gan, C. L., Prognostics and Health Management of Electronics: Fundamentals, Machine Learning, and the Internet of Things: John Wiley & Sons Ltd., pp. 731, *Life Cycle Reliability and Safety Engineering*, 9, 2, 225-226, 2020, ISBN: 9781119515326 (Print), 9781119515326 (Online).

Jardine, A.K., Lin, D., Banjevic, D., A review on machinery diagnostics and prognostics implementing condition-based maintenance. *Mech. Syst. Signal Process.*, 20, 7, 1483–1510, 2006.

Nguyen, D.N. and Zerhouni, N., Hybrid models for remaining useful life prediction using support vector regression and deep neural networks. *IEEE Trans. Ind. Electron.*, 65, 10, 7820–7828, 2018.

Rausand, M. and Hoyland, A., System reliability theory: models, statistical methods, and applications, vol. 396, John Wiley & Sons. 2003.

Xie, M. and Goh, T.N., Maintenance policies for stochastically deteriorating single-unit systems. *Eur. J. Oper. Res.*, 139, 3, 469–487, 2002.

10

Role of CBM and RLP in the Performance of the Manufacturing Industry

Harpreet Sharma[1]*, Chandan Deep Singh[2] and Kanwal Jit Singh[3]

[1]*Yadavindra Department of Engineering, Guru Kashi Campus, Talwandi Sabo, Punjab, India*
[2]*Punjabi University, Department of Mechanical Engineering, Patiala, India*
[3]*School of Agriculture Sciences and Engineering, Maharaja Ranjit Singh Punjab Technical University, Bathinda, India*

Abstract
In the case of a high-valuable asset, the maintenance phase requires heavy charges and more efforts because it has long usage life and any accident of an asset during this period causes catastrophic damage to an industry. Well-performed maintenance implies seeing as few corrective maintenance actions as possible while performing as little preventive maintenance as possible. This might seem as a utopia, but during the past decades, strategies and concepts have evolved for support. One of these is condition-based maintenance. In condition-based maintenance, critical item characteristics are monitored (through, for example, vibration or temperature monitoring) to gain early indications of an incipient failure. Therefore, the purpose of this chapter is to investigate how a condition-based maintenance approach can be implemented in an industrial setting, and to develop a method that can assist companies in their implementation efforts.

***Keywords*:** Condition-based maintenance (CBM), reliability-centered maintenance (RCM), predictive maintenance (PdM), asset health monitoring, operational efficiency, failure prevention, downtime reduction

**Corresponding author*: harpreetsharma3211@gmail.com

Chandan Deep Singh, Davinder Singh, Kanwal Jit Singh and Harleen Kaur (eds.) Condition-Based Maintenance and Residual Life Prediction, (191–206) © 2025 Scrivener Publishing LLC

10.1 Introduction

Increased productivity is a key issue for manufacturing companies to stay competitive on a global market. Success, and even survival, in manufacturing requires continuous development and improvement in the way products are being produced (Jackson and Petersson, 1999). Just-in-time, supply chain management, lean manufacturing, capacity assurance, flexible and agile manufacturing, to name a few, are strategies in which it is essential that production capacity is available to meet customer demand (Desirey and Dunn, 2000). Maintenance as a form of production support has thus become increasingly important to ensure equipment availability, quality products, on-time deliveries, and plant safety (Bevilacqua and Braglia, 2000). Even so, maintenance is still considered a cost center in many companies.

However, improper maintenance and unavailable equipment often limit the effectiveness of manufacturing. One-third of the total maintenance cost is spent unnecessarily because of circumstances such as bad planning, overtime costs, poor usage of work order systems, and limited or misuse of preventive maintenance (Bertolini *et al.*, 2004). There is also no doubt that maintenance is a costly support function. On a theoretical note, good maintenance has been defined as when very few corrective maintenance actions are undertaken and when as little preventive maintenance as possible is performed. This demands great skills in planning proper preventive maintenance intervals and tasks.

In general, maintenance is defined as all technical and managerial actions taken during usage period to maintain or restore the required functionality of a product or an asset. There have been various classifications of maintenance policies. These can be divided simply into breakdown maintenance and preventive maintenance. Another way to classify maintenance policy is into these three types: breakdown maintenance (corrective maintenance), preventive maintenance, and condition-based maintenance (CBM). In breakdown maintenance, the maintenance action is taken after some problems such as breakdowns in a product are found while preventive maintenance periodically checks a product with a certain interval to prevent the abnormality of the product. CBM may be similar to the preventive maintenance in the sense that its goal is to prevent product abnormality in advance before abnormality occurs (Ahmad and Kamaruddin, 2012). It focuses on the prediction of the degradation process of the product, which is based on the assumption that most abnormalities do not occur instantaneously, and usually, there are some kinds of degradation process from normal states to abnormalities. Hence, unlike breakdown maintenance and preventive maintenance, CBM focuses on not only fault detection and diagnostics of components but also degradation monitoring and failure prediction. Generally, CBM can be treated as a method used to reduce the uncertainty of maintenance activities and is carried out according to the requirements indicated by the equipment's condition (Prajapati *et al.*, 2012). Thus, CBM enables us to identify and solve problems in advance before product damage occurs. In industry systems, any product damage can lead to serious results. In this respect, CBM is quite an attractive method for the industry that operates high-valued assets.

Until now, it has been difficult to achieve effective maintenance operations because there is no information visibility during the product usage period. However, recently, with emerging technologies such as radio frequency identification (RFID), various sensors, micro-electro-mechanical system (MEMS), wireless telecommunication, and supervisory control and data acquisition (SCADA), product embedded information devices (PEID) are expected to be rapidly used for gathering and monitoring the status data of products during their usage period. Advancements in information technology have accelerated the growth in the CBM technology area by enabling network bandwidth, data collection and retrieval, data analysis, and decision support capabilities for large data sets of time series data (Quatrini *et al.*, 2020). Under the new environment, we can gather product status and usage data related to distributing routes, usage conditions, failure, maintenance or service events, and so on. These data enable us to diagnose the degradation status of the product in a precise way. Therefore, using this information gives us new challenging issues for improving the efficiency of product maintenance operations. We can diagnose product status, predict product abnormality, and execute proactive maintenance, i.e., do CBM (Rath *et al.*, 2022).

Since a critical failure or degradation of the product during its operation can seriously damage the belief of customers on the product's reliability, the maintenance enhancement for preventing this kind of failure or degradation in advance has precedence over any other things in a company. For this purpose, recently, many manufacturing companies are trying to adopt new technologies and get more accurate real-time information regarding product status during its usage period. As diverse information becomes available, the CBM approach to use them for preventing a critical failure or degradation in advance has been highlighted (uit het Broek *et al.*, 2021). Although most machine maintenance today is still either purely reactive (fixing or replacing equipment after it fails) or blindly proactive, world-class companies are moving forward toward 'predict-and-prevent' maintenance, which is very similar to the goal of CBM.

10.2 What is Condition-Based Maintenance (CBM)?

10.2.1 Definition:– What Does Condition-Based Maintenance (CBM) Mean?

Condition-based maintenance (CBM) is a maintenance strategy that monitors the real-time condition of an asset to determine what maintenance needs to be performed. Unlike preventive maintenance, which uses things like calendar-based maintenance or other means to determine when to schedule and perform maintenance, condition-based maintenance dictates that maintenance should only be done when these real-time indicators show irregularities or signs of decreasing performance (Liu *et al.*, 2022).

The goal of condition-based maintenance is to continuously monitor assets to spot impending bearing failure, so maintenance can be proactively scheduled before the failure occurs. The idea is that this real-time monitoring will give maintenance teams enough lead time before a failure occurs or performance drops below an optimal level.

Examples of condition-based maintenance:

- Monitoring the temperature of computers and machinery to prevent overheating or using smart HVAC units to control building temperature and save energy.
- Monitoring pressure in a water system to predict when a pipe could fail.
- Monitoring oil particles in construction or fleet vehicles.

10.2.2 CBM Typically Involves the Following Steps

Data collection: Various sensors, instruments, or monitoring systems are employed to collect data on the performance, health, and operating parameters of the equipment or system. This can include measurements such as temperature, vibration, pressure, fluid analysis, or electrical parameters.

Data analysis: The collected data are analyzed using various techniques such as statistical analysis, data mining, machine learning, or expert systems. The objective is to identify patterns, trends, anomalies, or deviations from normal operating conditions that may indicate a potential failure or degradation.

Condition assessment: Based on the data analysis, the condition of the equipment or system is assessed. This assessment can range from simple threshold-based indicators to more sophisticated models that consider multiple variables and historical data.

Maintenance decision-making: The condition assessment results are used to make maintenance decisions. If the analysis indicates that the equipment operates within acceptable parameters, no immediate maintenance action may be required. However, if a potential issue is detected, a maintenance plan can be initiated.
Maintenance actions: The maintenance actions can vary depending on the specific needs identified through CBM. It can include activities such as preventive maintenance, predictive maintenance, online monitoring, lubrication, calibration, part replacement, or adjustments to operating parameters.

10.3 Types of Condition-Based Maintenance

One of the biggest advantages of condition-based maintenance is that it is non-invasive, meaning real-time data are being collected while the machine is still running without adjusting the way it operates. You can choose to collect data at certain intervals or continuously through things like sensors, visual inspection or scheduled tests. Let us take a look at some of the most common types of condition-based monitoring techniques used in CBM.

Vibration analysis: This is defined as the process for measuring vibration levels and frequencies of machinery and using that information to analyze the health of machines and their components. Vibration analysis can help detect problems like imbalance, bearing failure, mechanical looseness, resonance, and bent shafts.

A simple example might look like this: Imagine you have an industrial fan. You remove one of the fan blades and start it up. As you might expect, the fan starts to vibrate due to the unbalanced fan wheel. This imbalanced force will happen one time per revolution of the fan, producing increased vibration signals. You could also have a damaged bearing track causing a bearing roller to generate vibration each time it contacts the spall. So, if three bearing rollers hit the spall per revolution, you would see a vibration signal three times the fan's running speed.

Infrared thermography: This is the process of using a thermal imager to detect radiation coming from an object, converting it to temperature and displaying an image of the temperature distribution in real time. Often utilized with a baseline image for comparison, infrared thermographic images can clearly and easily show when an asset is becoming overheated.

Infrared thermography is employed for monitoring the electrical and mechanical conditions of motors, inspecting bearings, and examining refractory insulation, as well as for checking gas, liquids, and sludge levels.

Infrared tools include spot infrared thermographers, which are used to measure thermal radiation on hard-to-reach assets or assets operating under extreme conditions; infrared scanning systems, which scan larger areas or objects on a conveyor belt; and infrared thermal imaging cameras, which measure temperature at multiple points across a large area and create two-dimensional thermographic images.

Ultrasonic analysis: This uses sound to identify potentially failing assets by detecting high-frequency sounds and converting them into audio and digital data. Data collection methods determine the types of detectable failure concerning ultrasonics. You can have either contact (structure-borne) or non-contact (airborne) methods. Contact methods are generally used for mechanical issues like bearing faults, lubrication problems, gear damage, and pump cavitation. All of these faults emit a high-frequency noise. Ultrasonic contact methods are also helpful for detecting electrical faults on motors, as loose or broken rotor bars can generate a high-frequency, rhythmic pattern. Lastly, failing steam traps may have steam constantly leaking past the internal seals, causing a rattle, which is picked up with ultrasound.

Non-contact methods (airborne) of ultrasonic measurements include pressure and vacuum leaks on compressed gas systems and a number of electrical applications. Air surveys use ultrasonics to detect leaks in compressed gas systems. Regarding the use of airborne ultrasonics for electrical systems, ultrasonic methods can detect arcing and corona when thermography cannot.

Oil analysis: This is a routine activity for analyzing oil health, contamination, and machine wear. An oil analysis program helps verify that a lubricated machine is operating as it should. Oil analysis checks the oil's fluid properties, answering questions like "Are the right additives active?" Have additives depleted? Is the viscosity where it needs to be? Oil analysis also looks to see whether there are destructive contaminants in the oil, and if so, helps narrow down the probable source. Lastly, oil analysis lets you analyze the presence of particles produced from mechanical wear, corrosion, or other machine surface degradation.

Electrical analysis: This is employed to examine the incoming power quality of assets using motor current readings from clamp-on ammeters to measure the current in a circuit. This makes it easier for maintenance personnel to see when an asset is getting an abnormal amount of electricity.

Pressure analysis: Maintaining the correct pressure within a piece of equipment to let fluid, gas, or air move through a pipeline or hydraulic hose properly is vital. Pressure analysis can continuously monitor pressure levels in real-time and alert to sudden drops or spikes, allowing maintenance personnel to respond to and fix issues before a more serious incident occurs.

10.4 When to Use Condition-Based Maintenance

Maintenance programs are diverse, and it is important that organizations select the right balance of different types of maintenance to be most effective. Condition-based maintenance has its strengths and is best for applications that require constant or regular monitoring of a factor that can reliably alert technicians to impending failures. These should be critical assets required for optimal production line performance or employee safety. The most common applications are as follows:

Temperature-sensitive assets: Temperature is a significant and easily measured indicator of equipment performance. It can be applied in a wide variety of situations. For example, ensuring a freezer remains within safe temperature ranges guarantees quality and safety in the food industry, keeping a motor within acceptable temperature ranges prevents overheating, and monitoring building temperatures can keep employees comfortable while conserving energy.

Pressure-sensitive applications: Sensors that can take continuous pressure readings greatly serve organizations that depend on assets using pressured air, water, or gas for their operations. Not only can tiny leaks be detected immediately using drops in pressure, but major damage and employee injury can be prevented if high-pressure situations can be mitigated as soon as they occur.

Oil-reliant equipment: Vehicles are the most common assets that benefit from oil analysis in condition-based maintenance. Just as your personal car requires regular oil changes, so do fleet equipment or other transport equipment. Instead of basing these oil changes on time or usage, analyzing the oil viscosity and particles can help better predict the life of the oil and when it requires replacement for optimal performance.

10.5 Steps to Take Before Implementing Condition-Based Maintenance

As discussed earlier, there are certain things you can do to get the most out of a condition-based maintenance plan.

Ensure a solid foundation: Condition-based maintenance goes hand-in-hand with reliability-centered maintenance (RCM) because RCM helps you identify potential problems with your assets and determine what you should do to ensure those assets continue producing at maximum capacity. Having a solid grasp of RCM processes helps focus your condition-based maintenance efforts where they need to be. Reliability experts agree that one of the biggest issues with adopting a condition-based maintenance program is the lack of understanding of RCM principles.

Include affected personnel: Once you have established that all maintenance personnel have the necessary skills, include them in the criticality analysis. Incorporating their input makes them active participants and enables them to use their RCM fundamentals effectively while contributing to the condition-based maintenance implementation. It also will help them to identify, mitigate, and eliminate failure modes.

Perform criticality analysis: As briefly mentioned earlier, a criticality assessment ensures your condition-based maintenance program is effective. Accurately identifying assets as critical, semi-critical, and non-critical can decrease unnecessary route-based maintenance. In other words, maintenance personnel will know which assets are the most critical and can perform checks on those assets first or more often than non-critical assets without making needless routes all over the plant.

Criticality analysis also helps determine which assets will benefit most from condition-based monitoring techniques like remote vibrations or acoustic sensors that produce real-time data that can be analyzed from another location. These more critical assets are referred to as "bad actors" or the repeat offenders in your asset lineup. These bad actors benefit more from continuous monitoring since they tend to have frequent issues. Keep in mind that once a criticality analysis is completed, it is not uncommon to find that the assets you once considered critical are not as critical as you previously thought.

Follow up: After you have finished your criticality assessment, it is a good idea to implement failure reporting, analysis, and corrective action system (FRACAS) to ensure your analysis is correct and the most critical assets are benefiting the most from your condition-based maintenance program.

10.6 Challenges of Condition-Based Maintenance

So, what is the catch? As with any process change or new process implementation, condition-based maintenance comes with some challenges.

Significant initial cost: The upfront costs associated with CBM tend to add up as you perform a criticality analysis and figure out where you need to place sensors. This can be even more costly if you must retrofit them on older assets. This is partly why your criticality analysis is so important because it determines which equipment will give the highest ROI. Newer or smaller plants might not have the onsite expertise to perform this kind of analysis, so it is wise to bring in an expert to conduct a failure mode and effects analysis (FMEA) and an RCM analysis, which will be an additional cost.

Also, choosing the proper sensor is critical. Consider factors like operating conditions, as sensors built to withstand harsh operating environments generally cost more.

Training: Now that you have sensors to provide real-time data and insight into the equipment's condition, you must have personnel who can analyze these data properly and quickly. For each fault detection or alert produced by a sensor, multiple questions arise. Does a part need to be replaced? Is the part in stock? How long do we have before the asset fails? Do we need the vendor to do the replacement?

Keep in mind that training is another expense and involves pulling operators and other maintenance staff away from their normal operating duties. Training also involves getting everyone on board with the change and effectively managing the change. As learned from the International Atomic Energy Agency (IAEA), this is one of the most difficult parts of implementing a condition-based maintenance program.

Operating conditions: The accuracy and performance of your sensors partly depend on the environment in which they are functioning. Harsh operating conditions can lead to malfunctioning or damaged sensors. For example, high heat and humidity can affect electronics, while corrosive chemicals can damage sensors and yield inaccurate readings.

Unpredictability: Unlike scheduled maintenance, maintenance work based on a condition-based monitoring program is unpredictable. For instance, you may perform maintenance when a sensor alerts you. This can cause irregularity in how costs appear in your budget. For example, if a handful of assets require maintenance at the same time, your maintenance team must be able to manage the repairs quickly.

Software requirements: Each installed sensor collects massive amounts of data continuously, so it is important to have a modern computerized maintenance management system (CMMS) or other software that can organize, track, collect, and analyze such data. Along with having the right software, you will need to consider hiring a third party to help analyze the results until your staff is fully trained.

Additionally, ensure your Wi-Fi connection can handle the amount of data being used and your cloud storage plan is large enough to hold the stored data.

10.7 Benefits of Condition-Based Maintenance

Implementing a condition-based maintenance component within your overall maintenance program can generate a great deal of benefits for your organization. Some key benefits include:

Increased reliability: When your assets can essentially "tell" you when they are about to fail, you will be able to experience much higher equipment reliability.

Decreased downtime: This, in turn, decreases downtime, which means your operators stay busy during their shifts and your equipment works at optimal levels.

Higher productivity: Long-term, this means higher productivity. More quality products can come off your lines, keeping customers happy and revenue climbing.

Reduced maintenance costs: Keeping your assets running smoothly is important, but too much unneeded maintenance can be costly. Condition-based maintenance means you are zeroing in on the problems that need to be fixed immediately, maximizing those maintenance dollars.

Rapid failure identification: Since potential failures can be triggered by temperature or pressure sensors, you will be notified quickly of problems.

Reduced unplanned failures: As a result, you can generate work orders for scheduled downtimes instead of shutting down a line for emergency maintenance.

Improved safety: Since condition-based maintenance can "catch" problems before they become significant issues, so you will be able to keep employees safer.

Increased asset lifespan: Small problems in equipment, left unchecked, can cause not only bigger problems but can result in further damage and larger, more expensive repairs. Condition-based maintenance can mean adding years of life to critical assets.

10.8 Role of Condition-Based Maintenance (CBM) on the Performance of the Manufacturing Industry

Condition-based maintenance (CBM) plays a crucial role in improving the performance of the manufacturing industry. CBM is an approach to maintenance that focuses on monitoring the condition of assets in real-time to make maintenance decisions based on their actual health status rather than predetermined schedules. Here are some ways CBM contributes to the performance of the manufacturing industry:

Improved equipment reliability: CBM allows for proactive maintenance by continuously monitoring the condition of the equipment. By detecting early signs of degradation or faults, maintenance actions can be taken before a failure occurs, reducing the likelihood of unplanned downtime. This improves equipment reliability and availability, minimizing disruptions to production schedules.

Increased equipment lifespan: By monitoring the condition of assets and addressing issues promptly, CBM helps extend the lifespan of the equipment. Regularly assessing the health of critical components, identifying potential failures, and taking appropriate maintenance actions can prevent further deterioration and costly replacements, leading to cost savings for the manufacturing industry.

Enhanced operational efficiency: CBM enables manufacturers to optimize maintenance activities by performing them when necessary rather than on a fixed schedule. This approach reduces unnecessary maintenance and minimizes production downtime caused by maintenance activities. As a result, overall operational efficiency is improved, leading to increased productivity and reduced costs.

Cost reduction: CBM helps manufacturing industries reduce maintenance costs in several ways. By adopting a condition-based approach, maintenance activities become more targeted and efficient, reducing the need for excessive maintenance. Additionally, proactive maintenance prevents major breakdowns and associated costly repairs or replacements. CBM also aids in optimizing inventory management by ensuring that spare parts are available when needed, minimizing inventory carrying costs.

Data-driven decision-making: CBM relies on data collected from sensors, monitoring systems, and other sources. By leveraging data analytics and machine learning techniques, manufacturers can gain valuable insights into asset performance, failure patterns, and optimal maintenance strategies. This data-driven decision-making process allows for continuous improvement, the identification of root causes, and the implementation of preventive measures to optimize the overall performance of the manufacturing industry.

Overall, CBM plays a pivotal role in the manufacturing industry by improving equipment reliability, increasing equipment lifespan, enhancing operational efficiency, reducing costs, and enabling data-driven decision-making. By shifting from reactive maintenance to a proactive and condition-based approach, manufacturers can achieve higher productivity, minimize downtime, and stay competitive in the market.

10.9 Residual Life Prediction

Residual life prediction, also known as remaining life estimation, is a field that deals with estimating the remaining useful life of a particular object or system. It is often used in engineering and maintenance applications to assess the health and longevity of various assets, such as machinery, structures, or electronic components.

Predicting residual life involves analyzing the historical data and current condition of the object or system to estimate how much longer it is expected to function properly before it fails or becomes unreliable. The goal is to optimize maintenance schedules, avoid unexpected failures, and reduce downtime and costs.

Several techniques are employed for residual life prediction, depending on the nature of the object being analyzed:

Statistical analysis: Statistical methods analyze historical failure data and use statistical models to estimate the remaining life. These models may include methods such as survival analysis, which considers the time to failure.

Prognostics and health management (PHM): PHM techniques integrate real-time monitoring data with predictive models to continuously assess the health of the system and predict remaining life. This approach is commonly used in complex systems like aircraft, power plants, or vehicles.

Physics-based models: For objects with well-understood physical behavior, physics-based models simulate the degradation and wear-out processes to estimate the remaining life. These models rely on knowledge of the object's materials, operating conditions, and stress factors.
Machine learning and artificial intelligence: Advanced machine learning algorithms, such as regression models, decision trees, neural networks, or support vector machines, can be trained on historical data to predict residual life. These models can capture complex patterns and dependencies in the data.

It is important to note that residual life prediction is not an exact science and can be challenging due to uncertainties and variations in operating conditions, maintenance practices, and other factors. The accuracy of predictions depends on the availability and quality of data, the chosen prediction method, and the specific application.

Moreover, it is crucial to regularly update and recalibrate prediction models as new data become available to improve accuracy and reliability.

10.10 Role of Residual Life Prediction on the Performance of the Manufacturing Industry

Residual life prediction plays a crucial role in the performance of the manufacturing industry by enabling proactive maintenance strategies and optimizing resource allocation. Here are some key ways in which residual life prediction impacts the industry's performance:

Predictive maintenance: Residual life prediction helps manufacturers anticipate the remaining useful life of equipment and machinery. By analyzing historical data, sensor readings, and other relevant factors, manufacturers can identify potential failures or performance degradation in advance. This allows them to schedule maintenance activities more efficiently, reducing unplanned downtime and improving overall equipment effectiveness (OEE).
Minimization of costly failures: By accurately predicting the residual life of critical components or systems, manufacturers can take preventive measures to avoid costly failures. They can proactively replace or repair parts before they reach the end of their life cycle, thereby reducing the risk of breakdowns that can result in production delays, quality issues, and expensive repairs.

Optimization of inventory management: Residual life prediction helps manufacturers optimize their inventory management practices. By knowing the remaining useful life of components or parts, they can adjust their stock levels accordingly. This prevents overstocking or understocking situations, reducing inventory carrying costs and ensuring the availability of critical spare parts when needed.

Resource planning and allocation: With accurate residual life predictions, manufacturers can better plan and allocate resources. They can optimize workforce scheduling, spare parts procurement, and maintenance activities based on the projected life cycle of equipment. This improves resource utilization, minimizes downtime, and enhances overall operational efficiency.

Enhanced equipment performance: Residual life prediction enables manufacturers to identify opportunities for performance improvement. By monitoring the health and degradation patterns of equipment, they can proactively identify inefficiencies, optimize settings, and implement corrective actions. This leads to improved equipment performance, increased production output, and better product quality.

Extended asset lifespan: Accurate residual life prediction can help extend the lifespan of manufacturing assets. By understanding the remaining useful life, manufacturers can implement appropriate maintenance and refurbishment strategies. This helps them maximize the value of their assets, delay capital investments, and achieve a higher return on investment (ROI) for their equipment.

Data-driven decision-making: Residual life prediction relies on data analysis techniques, such as machine learning and predictive analytics. By leveraging data from sensors, historical records, and maintenance logs, manufacturers can make data-driven decisions. These insights enable them to optimize their processes, identify recurring issues, and implement continuous improvement initiatives, ultimately enhancing overall performance.

In summary, residual life prediction plays a critical role in the performance of the manufacturing industry by enabling proactive maintenance, minimizing failures, optimizing inventory management, improving resource planning, enhancing equipment performance, extending asset lifespan, and facilitating data-driven decision-making. It helps manufacturers optimize their operations, reduce costs, increase productivity, and achieve competitive advantages in today's highly competitive market.

10.11 Conclusion

Both condition-based maintenance (CBM) and residual life prediction are valuable approaches that contribute to the performance and efficiency of various industries, including manufacturing.

CBM focuses on monitoring the real-time condition of assets, allowing for proactive maintenance based on the actual health status of the equipment. By adopting a condition-based approach, organizations can improve equipment reliability, increase equipment lifespan, enhance operational efficiency, and reduce maintenance costs. The ability to detect early signs of degradation or faults enables timely maintenance actions, minimizing unplanned downtime and disruptions to production schedules.

Residual life prediction, on the other hand, involves estimating the remaining useful life of an object or system. By analyzing historical data and current conditions, and utilizing techniques such as statistical analysis, PHM, physics-based models, or machine learning, organizations can make predictions about when assets are likely to fail or become unreliable. This information helps in optimizing maintenance schedules, avoiding unexpected failures, reducing downtime, and controlling maintenance costs.

Both CBM and residual life prediction complement each other in optimizing maintenance strategies. CBM provides real-time monitoring and proactive maintenance actions based on the condition of assets, while residual life prediction assists in making informed decisions about the timing of maintenance activities and resource allocation.

By integrating CBM and residual life prediction into maintenance practices, organizations can maximize asset performance, extend equipment lifespan, minimize downtime, reduce costs, and make data-driven decisions for maintenance planning and resource management.

Overall, these approaches play a vital role in improving operational efficiency, asset management, and competitiveness across various industries, enabling organizations to achieve higher productivity, reliability, and cost-effectiveness.

References

Ahmad, R. and Kamaruddin, S., An overview of time-based and condition-based maintenance in industrial application. *Comput. Ind. Eng.*, Elsevier, 63, 1, 135–149, 2012.

Bertolini, M., Bevilacqua, M., Braglia, M., Frosolini, M., An analytical method for maintenance outsourcing service selection. *Int. J. Qual. Reliab. Manage.*, Emerald Group Publishing Limited, 21, 7, 772–788, 2004.

Bevilacqua, M. and Braglia, M., The analytic hierarchy process applied to maintenance strategy selection. *Reliab. Eng. Syst. Saf.*, 70, 1, 71–83, Elsevier, 2000.

uit het Broek, M.A., Teunter, R.H., De Jonge, B., Veldman, J., Joint condition-based maintenance and condition-based production optimization. *Reliab. Eng. Syst. Saf.*, Elsevier, 214, 107743, 2021.

Desirey, S.T. and Dunn, R., Positioning maintenance as a competitive advantage. *Plant Eng.*, Reed Business Information 54, 5, 66–69, 2000.

Jackson, M. and Petersson, P., A model for development of future manufacturing systems. *CIRP Int. Semin. Manuf. Syst.*, 449–457, 1999.

Liu, Q., Ma, L., Wang, N., Chen, A., Jiang, Q., A condition-based maintenance model considering multiple maintenance effects on the dependent failure processes. *Reliab. Eng. Syst. Saf.*, Elsevier, 220, 108267, 2022.

Prajapati, A., Bechtel, J., Ganesan, S., Condition based maintenance: A survey. *J. Qual. Maint. Eng.*, Emerald Group Publishing Limited, 18, 4, 384–400, 2012.

Quatrini, E., Costantino, F., Di Gravio, G., Patriarca, R., Condition-based maintenance—an extensive literature review. *Machines*, 8, 2, 31, MDPI, 2020.

Rath, N., Mishra, R.K., Kushari, A., Aero engine health monitoring, diagnostics and prognostics for condition-based maintenance: An overview. *Int. J. Turbo Jet-Engines*, De Gruyter, 40, s1, s279–s292, 2024.

11

Impact of Competencies on Condition-Based Maintenance and Residual Life Prediction

Rajdeep Singh*, **Chandan Deep Singh and Talwinder Singh**

Department of Mechanical Engineering, Punjabi University, Patiala, India

Abstract

Competencies, condition-based maintenance (CBM), and residual life prediction (RLP) systems can all be used in a variety of industries, such as automotive, manufacturing, aviation, and healthcare. In this chapter, we'll take a quick look at CBM-RLP, competencies, and maintenance approaches. Manufacturing, defense, and other industries all use CBM-RLP and competencies extensively. To make the systems capable of maintenance, it involves a number of disciplines. Based on past and present system conditions, these disciplines aid in predicting future outcomes. The chapter offers a comprehensive analysis of CBM-RLP competencies that covers the practical applications of various approaches and technology and concludes by putting the CBM-RLP competencies system into practice.

Keywords: Residual, manufacturing, defence, RLP, CBM

11.1 Introduction

For a long time, maintaining any particular system's product lifetime has been an expensive and difficult task. Different strategies have been looked into to reduce system failures or remedy historical system failures, usually by increasing the frequency of maintenance operations. The challenges and inefficiencies of the previous generation of maintenance systems have been addressed by a series of improvement concepts. With the advancement of

**Corresponding author*: rajdeepsingh.87@gmail.com

Chandan Deep Singh, Davinder Singh, Kanwal Jit Singh and Harleen Kaur (eds.) Condition-Based Maintenance and Residual Life Prediction, (207–232) © 2025 Scrivener Publishing LLC

technology over the past few decades, maintenance tasks have undergone a significant evolution. Maintenance refers to a series of procedures or actions used to return a thing to a state where it can carry out its intended functions. Corrective maintenance (CM) and preventive maintenance (PM) are two primary categories for maintenance strategies. Corrective maintenance, commonly referred to as run-to-failure or reactive maintenance, is a technique used to return malfunctioning equipment to its intended function through repair or replacement. Due to unexpected failure, this technique results in high levels of machine downtime (production loss) and maintenance expenses (repair or replacement). The PM strategy is an alternative to the CM strategy (Prajapati, Bechtel and Ganesan, 2012).

11.1.1 Concept of Condition-Based Maintenance

The Rio Grande Railway Company first introduced the concept of CBM in the late 1940s under the name "predictive maintenance." The train firm employed CBM techniques to track variations in temperature and pressure measurements to identify coolant, oil, and fuel leaks in the engine. In terms of minimizing the effects of unforeseen failures and determining when to rectify a leak or replenish a coolant or oil sump, the CBM monitoring techniques were a remarkable success. This concept was very quickly recognized by the US Army, who eventually adopted it as a crucial maintenance tactic for supporting their military equipment (Ahmad and Kamaruddin, 2012).

Throughout the 1950s, 1960s, and early 1970s, CBM ideas and uses began to appear in a variety of industries. The key industries where CBM has been used and has demonstrated various benefits in terms of both efficiency and cost reductions are the automotive, aerospace, military, and manufacturing sectors. The US Department of Defense (Army, Air Force, Navy, and Marines) and firms like GM, Honda, GE, Digitech, Honeywell, and others are now investing in and collaborating on CBM technology applications. By allowing network bandwidth, data collecting and retrieval, data analysis, and decision support capabilities for massive data sets of time series data, advancements in information technology have added to and accelerated growth in the CBM technology sector (Jardine, Lin and Banjevic, 2006).

A vehicle's or any system's targeted data can provide deeper insights into system performance, system health, and failure root causes, as well as a prognosis of the system's or subsystem's remaining usable life. This has significant benefits for maintaining mission-critical systems used in the aerospace, military, maritime, automotive, industrial, and other sector.

With these useful applications and advantages, CBM has become a crucial competency area to integrate into a company's product line, whether it be for vehicles, aircraft, weaponry, or other items that require routine maintenance. By incorporating CBM technological enablers into their present and prospective system architectures, these sectors are concentrating on CBM principles and maintenance plans (Hao *et al.*, 2017).

The demand for CBM will rise as industries transition to a future where machines are unmanned and closely supervised by humans as they have been in the past. Examples of systems that potentially benefit greatly from CBM ideas and maintenance tactics include robotic systems, unmanned vehicles, windmill systems, manufacturing systems, and oil pumping systems. Businesses that use CBM as a maintenance approach could make considerable financial savings or increase operational efficiency. It might entail a reduction in workforce, a smaller supply footprint, cost savings from second- and third-order failure effects, less downtime, and other advantages relevant to their industry (Ellis, 2008).

Effective and efficient pre-determinations of when, where, what, how, and who will repair equipment systems are crucial for cost-effective maintenance, according to the examination of the aforementioned issues. An appropriate maintenance schedule should be created at an early stage, before serious equipment issues lead to deadly breakdowns. The number and cost of spare parts needed in the inventory for emergency repairs can be reduced by increasing the percentage of planned maintenance actions. Predictive choices based on condition-based fault diagnosis and the prediction of the trend of equipment deterioration are crucial for maintenance decision-making and planning procedures because maintenance must deal with the issues of intrinsic deterioration and random failure of equipment (Yam *et al.*, 2001).

The definition of condition monitoring is the gathering and analysis of pertinent equipment parameters with the aim of identifying equipment state variations from typical conditions and trends of the equipment's health. Equipment parameters are a collection of traits that reveal the state of an item. As long as the equipment is in good condition, these characteristics, such as vibrations and temperatures, often remain steady. Nevertheless, an anomaly in these traits can be a sign of a functional failure. Monitoring these prognostic metrics can provide information on impending equipment failure (de Jonge *et al.*, 2017).

A procedure or action known as corrective maintenance, sometimes known as the run-to-failure model, is necessary to stop a failure that has already happened or is now happening. To return the system to the way it was when it was brand-new, it can entail repairs, component replacements, or other actions. Regardless of the state of the system or subsystem,

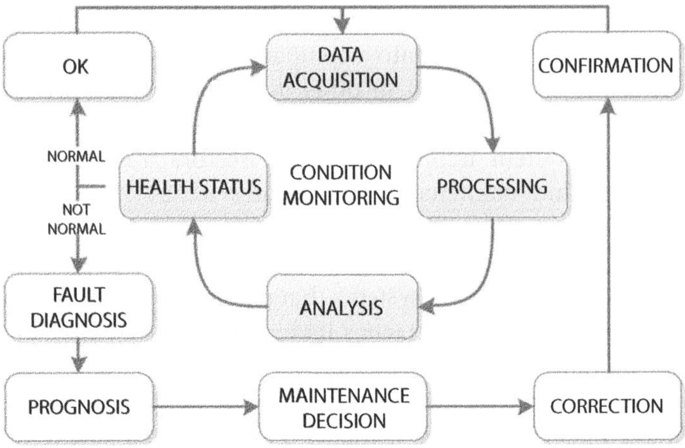

Figure 11.1 Condition-based maintenance process.

the schedule-based maintenance or preventive maintenance model is a maintenance philosophy based on predetermined intervals. For instance, even when a sizable percentage of the oil's life is still left, changing the oil in an automobile is routinely done either depending on mileage or time interval (Loué and Baronet, 2012). Figure 11.1 gives the process of CBM.

A strictly preventative or corrective maintenance strategy results in inefficient human resource utilization, system downtime, income loss, and other waste. For instance, most vehicles just change the oil periodically rather than according to how well it is doing at a particular moment in time. Without endangering the engine, the used oil might be used even more times. This may not be the best use of the available resources. Most machines and plants in manufacturing facilities need routine maintenance to prevent unanticipated repairs and downtime (Wang *et al.*, 2011).

11.1.2 Decision-Making in CBM

The sort of plant facilities that need to be covered, the accessibility of dependable surveillance equipment, investments in instrumentation and training, the amount of required staff, and running costs are only a few of the variables that influence what should be monitored. It should be emphasized that an impending functional failure may exhibit a variety of symptoms, each of which may become apparent at various phases of the unit's degeneration. For instance, altered vibration characteristics can serve as the first indication of a bearing failure. Increased particle content in lubricating oil samples, audible noise emission, and heat build-up on the bearing caps follow in that order.

Symptoms that appear in later stages give the occurrence of functional failure less time to prepare. When deciding on the most economical CBM process, it is important to take into account all of these symptoms and the condition-monitoring strategies that apply to them. Determine the time between inspections if equipment condition is not continually monitored so that timely action can be taken to either prevent functional failures or to lessen the effects of those that cannot be avoided. Over inspection will occur if this interval is too short (You and Meng, 2011).

Both static and dynamic warning limitations may be used in a condition-based maintenance program as catalysts for replacement and diagnostic action. The predetermined thresholds of the measured data are known as static limits. Examples of static limits for vibration monitoring include the ISO vibration severity charts. Even though CBM activities that are triggered by static limitations are simpler to manage, they lack the diagnostic capability to foretell when the alert will be met. On the other hand, dynamic limits are employed to track the pace at which the measured parameter is changing. When using dynamic warning limits, a CBM procedure prioritizes the rate of change of the monitored parameters over their actual value. Using dynamic limitations will undoubtedly improve a CBM procedure's automatic diagnostic capabilities (Jaske, 1985).

11.1.2.1 CBM Decision Models

The topic of inspection is one of the main classes of decision problems in CBM. It deals with circumstances when inspection duties are carried out on equipment at regular intervals to ascertain its actual condition. Two decisions will be made based on the findings of each inspection: (1) what maintenance action to take—to replace or repair the system to a certain state, or to leave it in its current condition; and (2) when the next inspection is to occur. Pure checking, shortened checking, and checking with monitoring are the three inspection policies that are taken into consideration. These policies presume discrete deterioration levels and base the determination of the next inspection time on the system's most recent state. When an object is discovered to be in an undefeated state, the next checking is always planned in accordance with the pure checking policy when an item is found to be in an unmatched condition. The item will be replaced when it has deteriorated to one of its late states, according to the shortened checking policy. The third policy is comparable to the shortened checking policy with the exception that when the item is discovered in a late stage, a continuous monitoring operation is started (Ray *et al.*, 2000).

These models' primary goal is to reduce the long-term estimated cost per unit of time. It is not specified how to find the ideal answer. To address issues like incomplete inspections, dangerous inspections where each inspection may result in an immediate failure or increase the failure rate, and the inspection of a stand-by system, other varieties of inspection models have also been developed.

11.1.2.2 Proportional Hazards Modelling

According to the traditional method, an item's failure risk is only based on its age or usage. It is acknowledged that equipment condition monitoring techniques, such as vibration monitoring and spectrographic analysis of oil samples, also offer a wealth of data that can increase the accuracy of the assessment of such risk. The approach for this evaluation is provided by proportional hazards modelling (PHM), a multivariate regression analysis technique first proposed. The PHM technique was first mostly used in the medical industry. The open literature first documented attempts to apply the technique to evaluate the reliability of airplane, marine, and locomotive engines in (Hao *et al.*, 2017).

The condition-based maintenance decision process can benefit from the additional data collected from condition monitoring jobs and analyzed by the PHM technique. To make the best replacement decisions, the study offers a control theory approach that includes PHM risk assessment in the cost consideration. The PHM procedure is used to assess the level of deterioration in that method, and the transition between levels is discretized and modelled as a semi-Markov process. Inspections take place at regular intervals. At each inspection, a choice is made between replacing the item right away and planning to keep using it for another x inspection intervals. Both the state and the age affect the value (Zhou *et al.*, 2014).

11.2 Application Areas of CBM

Condition-based maintenance (CBM) is used in a wide range of industries, including manufacturing facilities, the process sector, ground vehicles for the military, navy, and air force, IT infrastructure, and commercial vehicles aviation/aircraft. We gave the IT infrastructure some basic guidance on engine oil, vibration analysis, and temperature analysis. These are a few fields in which CBM is applied. CBM is being used in the highly targeted realm of manufacturing (Nurach *et al.*, 2012).

11.2.1 Automobile Sector

The engine oil used in military trucks, commercial trucks, and other engines that consume engine oil is the subject of this section's study. In typical service conditions, it is customary to change the oil every 3,000 or 7,500 miles or at any other interval, depending on the manufacturer. It was covered in earlier parts; however, it is possible that the oil quality did not go below the necessary level. It might continue to serve the engine for a few hundred miles without having an impact on performance. Currently, some high-tech cars come with sensors that monitor the oil's condition, alerting the driver to change the oil only when the quality drops below the required level rather than at predetermined intervals (Banjevic *et al.*, 2001).

The reasons why oil deteriorates over time can vary. During engine operation, engine oil mostly picks up some impurities. By extracting some samples from the engine, it may be examined. These substances, such as antifreeze, fuel, or dust, cause the engine oil quality to be much worse than it should be under regular operation.

11.2.2 IT Infrastructure

One of the major problems with IT infrastructure is maintenance. The information technology revolution has elevated IT infrastructure to the same level of significance as manufacturing plant machinery. Today, hard disks are used to store essential data in electronic form. Large data centers are being used by multinational corporations to store electronic data. The biggest loss for any large company could result from a single point of failure in the servers. These problems raise important challenges about how to maintain infrastructures safely in order to deliver high-quality services and lower the likelihood of data loss. Disk failures are caused by many factors (Bengtsson, 2004).

11.2.3 Process/Manufacturing Industry

Production plant maintenance is typically carried out after a failure. Depending on how bad the defective device is, it might be restored or replaced. The plant could experience prolonged downtime as a result of these unanticipated failures. Vibration sensors provide a wide range of failure analysis services. Any vital component, including motors, fans, gearboxes, pumps, vibratory conveyors, bearings, and other rotating mediums, can have a vibration sensor put on it. CBM is an alternative strategy that reduces downtime significantly by anticipating failure based on vibration analysis.

It is able to forecast how long the system will last. An fast fourier transform (FFT) analysis of the vibration of a particular system component can be used to quickly identify changes in vibration patterns and deterioration over time (Elena Dragomir *et al.*, 2009).

Although CBM is a very effective and beneficial maintenance strategy, its initial deployment costs might be rather high. Despite the initial deployment's high cost, it can be quite successful in achieving long-term maintenance and supportability objectives. Future systems are extremely sensor-equipped for a variety of uses. Within an existing infrastructure, a CBM application can be built more easily and with fewer adjustments. Data transfer across wired or wireless networks to maintenance management systems and enterprise resource planning systems is required, along with the monitoring of some specific components (Chen and Trivedi, 2002).

11.3 Residual Life Prediction

A crucial component of condition-based maintenance (CBM) decision-making is residual life prediction. Our goal is to create a residual life prediction model for the use of CBM decision-making in a situation with little or no information. In operational contexts, CBM techniques may be abused or applied ineffectively. The training and parameterization of the residual life prediction model require historical data on plant failures and CBM monitoring records, which will increase the effectiveness and efficiency of CBM (Huang *et al.*, 2007).

To construct and initialize the maintenance model, complete historical data sets of CM information with related failure times are, however, infrequently available, and the expenses of gathering adequate data can be significant. We are striving to solve these practical challenges with the research presented in this study to create a residual life prediction model for CBM that is as reliable as possible for CM and prognostic applications in the future. The kind of CM data that we take into account is of a "direct" kind, which applies when the measurement or observation directly connects to the status or condition of the plant, such as when the goal is to forecast how the degradation will progress. Direct monitoring is the term used for this. To specify the first passing time for the deterioration process to touch the line or zone in this situation, we require a threshold line or zone. The first hitting time measures how long it takes from the moment a CM reading is taken until the degradation signal first reaches the threshold line or zone (Boikov, 2017).

In the past few years, both academic and practical interests have been focused on the use of measured condition or degradation information to anticipate the residual life of important plant components. The ability of condition monitoring (CM) information to alert the plant manager of an impending breakdown allows condition monitoring (CM) strategies to enhance plant performance in terms of dependability, availability, or cost. If the residual life can be estimated with accuracy, timely corrective measures can be performed to stop the probable failure from occurring. The effectiveness of CM, among other things, depends on the ability to foresee the timing of an impending failure, or residual life, to use a more accurate term (Gebraeel *et al.*, 2004).

Here, the plant item could be a component or a system that is subject to CM and has information on its rate of degradation. In this paper, the term "CM information" refers to this degradation information. The availability of such a residual life projection does give the maintenance manager a tool for assurance that it will be easier to decide whether to maintain or replace the concerned plant using the measured CM data.

The residual life, however, is normally random unless there are unique circumstances where a prediction model based on physics may be created. Most of the time, we must use a distribution to predict the residual life probabilistically. This brings us to the main topic of this work, which is the probabilistic estimation of the remaining life of a monitored plant utilizing current state-of-the-art CM data. The well-known Brownian motion model with a drifting parameter is used in the development of the prediction model. However, this paper makes two original contributions. First, we make the Brownian motion model's drifting parameter adaptive, meaning that it will be updated as soon as a new observation of the CM data becomes available (Herzog *et al.*, 2009).

The conventional Brownian motion-based model, which has a fixed drifting parameter, has been significantly modified by this. The prognosis is based on the entire history of an item's monitored CM information up to this point using the proposed model, an aspect that is frequently overlooked in prognostic literature. The use of a threshold distribution rather than a single threshold line is the second contribution. We suggest the usage of a threshold zone represented by a distribution because it is frequently quite challenging to define a single threshold line in practice.

Many businesses must find a way to securely extend the operating life of existing plants or critical components because they cannot afford to replace them given the present economic and political context. For them, accurate techniques for estimating residual life are essential. When used as a foundation for prompt, secure, and cost-effective replacement or repair of

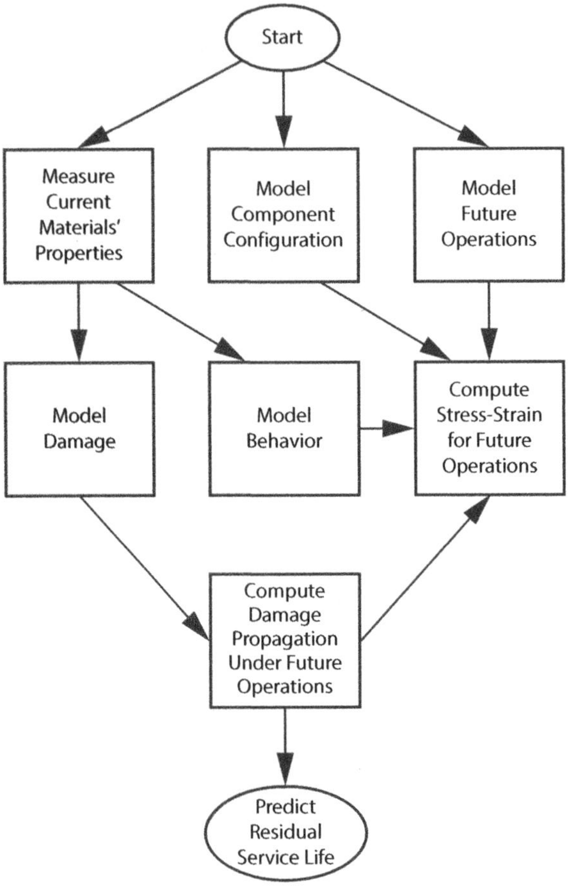

Figure 11.2 Residual life prediction framework.

crucial parts, accurate, trustworthy projections of remaining life can help to maximize a plant's utility. Figure 11.2 shows the framework of residual life prediction (Herzog *et al.*, 2009).

11.3.1 Technical Approach

A multidisciplinary technical approach is necessary for an accurate assessment of remaining life. The overall process and important components for residual-life prediction are shown in Figure 11.2. Depending on the particular application for which residual life is being examined, each of the factors may have varied degrees of complexity. First, the material's present condition, including its state of damage and current physical and mechanical qualities, is measured using one method or a combination of procedures. When the data are quantified, they serve as the foundation for modeling constitutive behavior as well as the future evolution of damage. The stress-strain environment history at crucial places in a component is estimated using the latter model, models of component configuration, and expected future operation history. The component's remaining life is then estimated using this result along with the damage model's output (Pawar and Ganguli, 2007).

By estimating the material's degradation from its original values as new material and its operation history, one might try to "measure" the material's damage condition and attributes. However, as noted by researchers, the application of this strategy is unclear and frequently results in wildly erroneous projections of remaining life. Therefore, the best course of action is to use some direct way to measure the material's qualities in its current state.

Non-destructive methods are preferred since the component is not harmed by them. Traditional non-destructive techniques have not been developed or calibrated to measure more subtle changes in material properties, such as creep voids or precipitate coarsening, and are typically confined to finding fractures and faults. Plastic replication is gaining popularity and has been proven to be a useful field tool for detecting minute changes, although its usage is restricted to accessible surface areas. Miniature specimens are being designed and tested for destructive testing since just a small amount of material needs to be taken out. Therefore, testing on small specimens is semi-destructive since the component can be used without suffering functional loss (Mohanty *et al.*, 2009).

Tool wear and the deterioration of product quality typically demonstrate an interactive relationship in multistage manufacturing processes (MMPs). On the one hand, the quality of the objects produced at a stage may suffer due to tool wear. On the other hand, the deterioration of quality at a previous stage may result in a change in operational circumstances and thus have an impact on tool wear at a later stage. To correctly anticipate the residual life distribution of MMPs, which will aid condition-based maintenance and tool inventory management, this relationship needs to be taken into account. In this research, we suggest an interaction model that makes use of a stochastic differential equation model to capture the impact of tool wear on quality degradation and a linear model to describe the impact of tool wear on quality degradation (Liu *et al.*, 2017).

Predictive maintenance's major goal is to keep track of the machine's component failures and operating conditions. The condition of the equipment should be constantly checked if a fault has been discovered in one of the components to prevent an unplanned failure. Some potential future critical symptoms related to the degeneration of the machine can be predicted by tracking the existing working circumstances of the malfunctioning machine. When a condition monitoring system that uses predictive analytics is installed, an advanced alarm can be issued when the anticipated value for the fault symptom falls within an alarm range (Dongshan, 2014).

11.3.2 Future Needs and Critical Issues

Due to the limitations of current residual-life evaluation methodologies, specialized procedures should only be used in certain circumstances where they have been validated. Over the next five years, there should be an increase in the amount of experimental data, damage models, computational software, recommendations, and application experience. This should result in an increase in the quantity and scope of standardized procedures for life prediction and damage assessment. Therefore, five years from now, more comprehensive general purpose methods for estimating the remaining useful life of pressure vessels and piping systems should be accessible. The following ten areas require technological growth and progress to ensure that truly general-purpose instruments do become available:

- A process for prioritizing and selecting certain components to be examined as well as the methods to be employed in doing so, as well as the costs and advantages of each strategy as it relates to the overall system of equipment.
- Probabilistic approaches to account for variances and uncertainties in material parameters, service loadings, analytical methods, experimental observations, and component configurations in creep and creep-fatigue life evaluation.
- A well-established database of properties of materials that have been in use, which working engineers can simply access and use. The database ought to be automated and accessible in a format that can be used with personal computers.
- Historical information on the equipment currently in the plant, such as as-built measurements, materials, fabrication techniques, particularly welding techniques, operational loadings, such as pressure, mechanical vibrations, and cyclic fluctuations, and environmental exposures, such as temperatures, gases, and liquids.

- Consensus codes and standards for component recertification and plant life extension that will assist users of life assessment technologies in addressing liabilities, risks, and insurance-related concerns.
- Damage accumulation models and life-prediction techniques for the onset of creep-fatigue, the growth of creep and creep-fatigue cracks, and thermal-mechanical fatigue (the majority of current models and techniques have only been created and validated for isothermal conditions, despite the fact that the majority of real-world applications involve thermal cycling).
- Generic stress studies that can be made simpler for use with personal computers, consider the effects of residual stresses, and use the emerging inelastic fracture-mechanics approach.
- Validated and demonstrated common testing methods for describing the mechanical characteristics and microstructure of materials exposed to service, particularly for field applications of plastic replication and small specimen techniques.
- Non-destructive techniques for measuring and tracking the mechanical characteristics and microstructural features of materials exposed to service.
- Pressure vessel and pipe system repair and rejuvenation procedures that use the outcomes of the life-assessment methodology for procedure selection and certification (Tao *et al.*, 2017).

It is imperative that future work focus on meeting each of the first five needs mentioned above. A user will be able to use the techniques best suited to their individual application by using an integrated systematic assessment procedure rather than taking a "hit-or-miss" approach. Deterministic models cannot produce a meaningful assessment of residual life, but probabilistic techniques can. All engineers will have access to a global resource through an established reference database when using residual-life evaluation techniques. As new information becomes available, the database needs to be updated and maintained. All companies who intend to apply residual life prediction technologies must gather, synthesize, and store historical data on as-built configurations and previous operations. These data should be stored in a standard format and, where practical, distributed among user groups with similar issues and interests. The safe implementation of life assessment and plant life extension requires the establishment of agreed norms and standards (Ilchenko *et al.*, 2012).

11.4 Competency Framework

A capability or aptitude is what is meant by a competency. It is a collection of connected but dissimilar behavioral patterns arranged around a central idea that we refer to as the "intent." The actions are different expressions of the intent that are appropriate for different circumstances or moments. For instance, two behaviors include listening to someone and asking them questions. These actions might be displayed by a person for a variety of causes or intentional purposes. To earn favor with another person or to appear interested, one can converse with them and listen to what they have to say. The main goal is to comprehend the individual. On the other hand, the earlier rationale for the interrogation was to gain favour or influence in the subject's eyes—a measure of what we could call a show of influence. Similar to this, self-insight and self-understanding are the basic goals of a more nuanced competency like emotional self-awareness (Bertoncelj, 2010).

Predicting effectiveness provides the framework for determining which actions and intentions are pertinent in a given circumstance. It involves connecting various behaviors that are thought of as alternate manifestations of the same underlying construct to develop a specific competency. However, these behaviors are initially grouped together based on how comparable the outcomes are when these actions are used in social or professional contexts. The organization's culture, climate, structure, systems, industry maturity, strategic positioning within it, and elements of the local economic, political, social, environmental, and religious climate are all predicted to have significant effects on how competencies are demonstrated and/or how jobs and roles are designed. The first four articles included in this special issue are investigations on the validity of competencies in four distinct professional occupations (Kaane, 2006).

Since David McClelland initially suggested competency-based human resources as a crucial performance differentiator, the idea has evolved from a novel strategy to a widespread practice. Nowadays, competency-based human resource management is used in practically every company with 300 or more employees. Competency evaluation and development are now practiced around the world by major consulting organizations like The Hay Group, Development Dimensions International, and Personnel Decisions Incorporated, as well as by thousands of small consulting businesses and

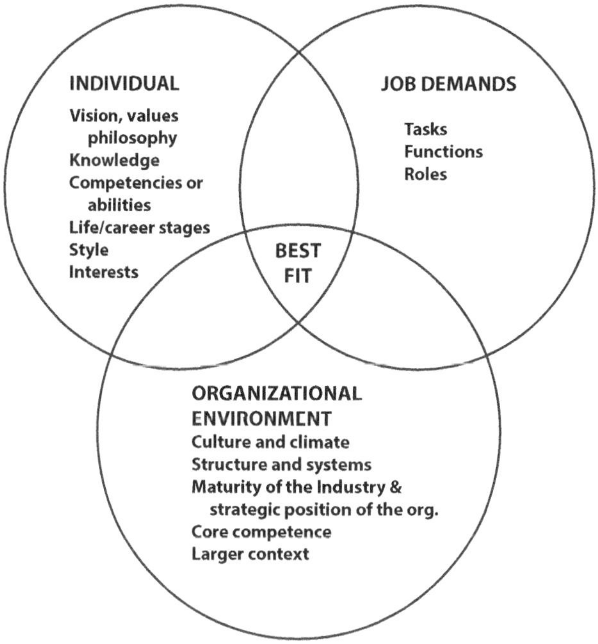

Figure 11.3 Competency framework.

freelance consultants. Yet, the literature on academic and applied research has lagged behind application. As a result, many academics and some professionals continue to be skeptical, and less information than necessary is provided to practitioners by ongoing research (Figure 11.3) (Yamani, 2021).

The Journal of Management Development's special issue on competen cies updates our knowledge of skills, how they affect performance, and how they are created. We include four studies that demonstrate the connection between performance in a variety of occupations, including bank executives, administrators of public schools, R&D managers, and military pilots, and demonstrated emotional, social, and cognitive intelligence competencies.

11.4.1 Competency Work Areas

A comprehensive understanding of emotional, social, and cognitive intelligence abilities provides more than just a useful foundation for outlining people's character traits. It provides a conceptual framework for the arrangement of personality and connects it to a philosophy of action and job performance. According to Goleman, "emotional competence" refers to a "learned capability based on emotional intelligence that culminates in great performance at work." In other words, a competency is an "underlying personality trait that promotes or results in effective or exceptional performance" (Momme, 2002).

- An emotional intelligence competency is the capacity to identify, comprehend, and make use of emotional information about oneself to achieve or cause successful or outstanding performance.
- A social intelligence competency is the capacity to identify, comprehend, and apply emotional data about people in a way that promotes or results in efficient or superior performance.
- A cognitive intelligence competency is the capacity to understand or evaluate information and situations that result in or cause effective or superior performance.

The inclination to think that more successful people possess the necessary components for success, if described as a single construct, invites the attribution of a halo effect. For instance, individual A is smart, savvy, and stylish, and because of these traits, she is effective. The problem of choosing the optimal level of detail in identifying constructs to use while developing a personality theory may eventually come down to choosing the best "focal point" with which to view something. We think that the distinct abilities and clusters are the most useful focal points for describing and researching performance (Sehr, 2005).

The personality theory that this technique is founded on informs how a skill is specified. According to the study's initial description, a person's unconscious motivations, self-schema, and observed behavioral patterns are connected to a theory of personality. This method was proposed by researchers as an integrated system diagram with concentric circles and the person's unconscious motives and characteristic dispositions in the center. The next expanding circle of the individual's values and self-image was influenced by these and was influenced by them. The skill level was indicated by a label around the circle. It was surrounded by a circle of observable, particular behaviors (Hameed Al Taee and Dina Fadhil Jihad Alwaely, 2012).

This conception of personality necessitates a wider-ranging viewpoint than is typically used. A more complete understanding of humans is revealed when the physiological level is combined with the psychological and behavioral levels. Due to the varied character of the several fields researching different aspects of this model, evidence of the causal sequence predicted in this personality theory is just slowly developing.

11.4.2 Competency Effect on CMB and RLP

There is a spectrum of proficiency, from novice to expert. Assessment metrics should be able to demonstrate a growth in competence rather than just demonstrating its lack. Different baselines or beginning points among learners are made possible by a range of skill levels. This range also enables students to find additional areas for learning and growth as they learn and to advance depending on increasing competence. To demonstrate the learner's capacity to comprehend and analyze the scenario, as well as to act in response to a particular circumstance, competencies should be tested in open and complicated situations. Working with cases in agency settings offers such open, complex systems in social work (Guo, 2000).

Valid assessment tools must be directly related to a particular competency or its subcomponents. The measure must be clearly and explicitly tied to the competency being evaluated. Measures of competency should, at the very least, be face valid in terms of research. For instance, observation or monitoring of such work should serve as a basis for determining a person's level of competence in providing social work services to a variety of populations. As appropriate and available, evaluations from other sources, such as clients, coworkers, and other service providers, may be included. It does not matter if the feedback is formal or informal, qualitative or quantitative, it just needs to be pertinent to the competency being evaluated (Manna *et al.*, 2016).

A important first step is to clearly connect the exam's content to the competency being evaluated. The policy must be evaluated contextually. Students could point out the policy's various advantages and drawbacks. The policy's many effects on individuals, families, groups, organizations, and communities may also be noted by the students. As part of this exam, the student must provide knowledge and proficiency in policy assessment, as well as provide some framing judgments. Value and ethical difficulties could surface in the student's evaluation of the policy's advantages and disadvantages. Various outcomes for different learners are very likely. Figure 11.4 shows the relationship among condition-based maintenance, residual life prediction, and competencies (Jain, 2013).

Capstone projects offer an additional method for evaluating the performance of complicated learners. Practice projects, theses, or other varied learning opportunities may be a part of capstone projects. A capstone project should demonstrate several distinct competencies or their constituent parts.

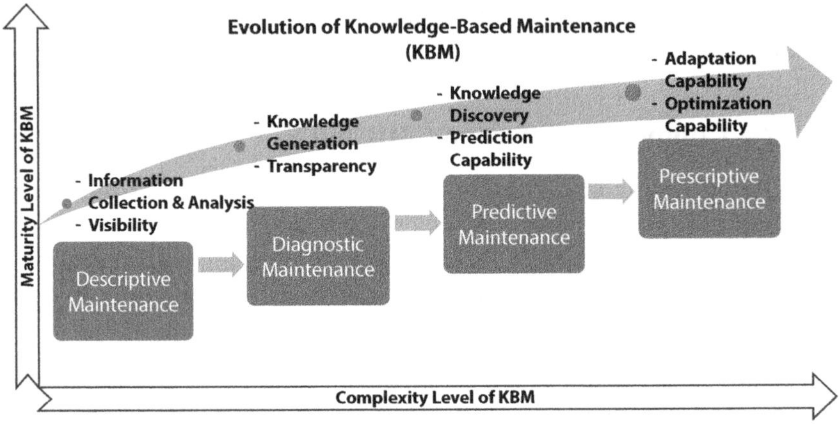

Figure 11.4 Competencies (capabilities) and maintenance relationship.

Measures for evaluating capstones should specify each ability and offer precise guidelines for evaluation. For instance, an assessment question on how practice experience is applied to inform scientific inquiry could be part of a thesis research project. It should be clear what criteria will be used to evaluate this item, maybe on an evaluation form. The form could include a Likert scale, a similar set of progressive but descriptive points, or categories like "omitted," "minimally addressed," and "completely explained."

The goal of this article was to establish skills and investigate how to evaluate them. However, choosing which specific competencies to investigate becomes a vital component of establishing a profession if competencies are to function as markers of professional performance. Although competencies can be thought of as including knowledge, values, and skill components, it is important to clarify precisely what they address. It is important to make a distinction between foundational and advanced or specialized outcomes. A profession's success depends on having a suitable, extensive, and adaptable list of skills and the practice behaviors that make up each one. Such lists require careful and meticulous definition (Hameed Al Taee and Dina Fadhil Jihad Alwaely, 2012).

The process of creating a list of skills should involve a variety of professional social work organizations and provide time for input from all social workers. Making the assessment of capabilities as meaningful and useful as possible depends on carefully and completely establishing the list of basic competencies at both the foundational and the advanced or specialized levels. This is a noble and significant activity.

It can be difficult for employers to determine the optimum use of certain abilities when managing performance improvements. They must also ensure that people have the necessary support to implement these abilities successfully. Our findings indicate clear tendencies across the employers in our case study for use, but unfulfilled needs are also highlighted by resounding signals from their users (Sigh *et al.*, 2008).

The effectiveness of human resource (HR) policies and processes depends on the ability to accurately, consistently, and fairly assess and quantify competencies as their use grows. Concerns about how organizations handle diversity and ideas of equality and fairness have also been high on the HR agenda. Competencies may be used to enhance the uniformity and fairness of the evaluation process since they offer a consistent vocabulary to discuss performance and explain the behaviors required. In fact, some employers implement them in the hope that they will increase evaluation consistency (Fouad *et al.*, 2009).

Another contentious issue is whether or not to relate pay to competency. Understandably, employers who have successfully employed competencies for development are hesitant to make such a connection for fear that it may reduce the advantages they obtain from creating them. Using skills to reward performance has been regarded as conflicting with developing them as it may decrease employees' willingness to report growth requirements (Strebler and Bevan, 1996). The discussion has centered more on our inability to quantify individual competency performance with accuracy (Hulla *et al.*, 2019).

Our first research efforts were concentrated on the procedures and problems associated with evaluating and measuring abilities in the context of performance reviews. The organizational context in which the evaluation and measurement procedure is used, however, determines its effectiveness. Given the potential problems we have discussed with the use of competencies, it was critical to determine how much of the problems with assessing and measuring competencies were caused by the competencies themselves and how much were due to other situational variables that were having an impact on them, such as the culture of the organization, current HR practices and policies, boss-subordinate relationships, purpose and outcome, etc. The area needs to be thoroughly mapped out to identify problems and best practices (Bhamra *et al.*, 2011).

Moreover, the experience of the organizations who had tried to introduce skills but had failed suggests that it may be more a matter of the right amount, done correctly, and at the right time. The organization, which was making a second attempt at implementing competencies, felt that ensuring user ownership was the way to go. Initially, managers had not been very supportive of competencies. This time, every manager received a letter asking for their feedback on the situation. A significant exercise in consultation was being carried out by HR employees with the assistance of occupational psychologists serving as internal consultants.

All businesses offered their staff members policies. The majority were in folders and ranged in length and level of detail. However, it appears that little instruction was offered on how to use skills because the majority of competency frameworks tended to be integrated with the performance evaluation system. Additionally, there was scant evidence of extensive training exercises conducted by organizations. In fact, some of the managers we spoke with in one organization had complained that they had only received a two-hour lecture on how to use the new competency-based appraisal system, despite the fact that HR professionals had spent many months designing it. However, one organization ran lectures that they referred to as "career fairs" (Vinet and Zhedanov, 2011).

The respondents who work in technical and specialized fields are the least satisfied with how well the competency framework matches the abilities they need for their jobs. As a result, it is suggested that technical and specialized personnel should be involved in the creation of competency frameworks and that employers acknowledge the possibility that general competency frameworks may need to be complemented by technical and specialized abilities.

The case study organizations' wage rises. Although competency performance is evaluated, the majority of the organizations questioned do not directly link these evaluations to wage increases. This may not come as a surprise given the lack of a pay-link between competencies and organizations that use them. It implies that the case study organizations are a good representation of how competencies are used by employers nowadays.

Competencies, however, are also the ones that are least frequently utilized to specify job criteria. It appears to confirm that requirements for competency performance are more driven by personal performance than by the tasks to be carried out, given that businesses are shifting away from defined work requirements to more flexible job roles. It illustrates how skills are used to define aspects of job performance that are not adequately captured by simply listing the duties and responsibilities of a job or job role. However, controlling employee performance without providing them with clear signals of what they must perform may offer issues (Singh and Khamba, 2016).

Comparing whether consumers' perceptions of the usage of competencies align with what their organizations claimed they were currently utilizing them for proved to be an intriguing exercise. For instance, one organization informed us that competencies were not evaluated for performance reviews and that they were not currently associated with remuneration. A sizable portion of their respondents have really used skills to evaluate work performance and, to a significant part, determine wage and pay raises. Other organizations also faced a mismatch, albeit one that was less visible. Disparities between recommended and actual practices further highlight the necessity of taking into account the application of competencies from the perspective of users, even though this may easily reflect different practices used by various parts of the organization or changes in usage over time (Miranda *et al.*, 2021).

Although respondents disputed that "competencies have helped them comprehend what other people's occupations entail," it is interesting to note that competencies are regarded to deliver clearer messages regarding the essential behaviors of employees. We found no difference between organizations utilizing a more specific framework, such as analytical competencies, although this may be because some organizations were employing broad competency frameworks. Making professional options naturally requires having a thorough understanding of other people's employment, which can also help with perceptions of internal equity.

These results demonstrate that using competencies to determine salary and pay increases "to a great extent" has the effect of increasing respondents' satisfaction with comprehending how competencies relate to reward and with reaching an understanding of their performance level with managers/direct reports. This may not come as a surprise given that connecting competencies to pay may emphasize the need for managers to be more transparent about these connections. Given that rewarding talents is seen to interfere with growing them, it could appear counter-intuitive (Shah, 2016).

Overall, survey respondents did not feel that competency-based evaluation required too much time to be effective in practice, which suggests that with practice, the process may become simpler to apply or that these organizations had been more successful in putting their approaches into practice. We believe both of these factors may be relevant. However, this initial assessment stage should not be disregarded because competencies vary and new employees need to get used to the approach. While acknowledging the present level of competency performance is a prerequisite for implementing competencies, the initial assessment stage should not have much weight because it is retrospective. For the following round of performance reviews, it is best to develop competency targets during this first step. We found little indication in the published recommendations that the initial evaluation stage and the concerns surrounding it were being highlighted (Banerjee, 2003).

However, across the case study organizations, interpersonal skills and soft and behavioral competencies were seen as the most difficult to assess and develop. Again, various explanations are pointed out in respondents' comments. Soft skills can be expressed in a variety of ways, such as the following: "Sensitivity and flexibility are rather woolly and serve little purpose other than as a hook for criticisms which could be made more informally"; "I cannot see how assertiveness differs from self-confidence"; and "Individuals relate to others in different ways because of the context of the situation they are in" (Shaheen *et al.*, 2019).

Interpersonal skills have gained importance as organizations seek to promote empowerment. It goes without saying that hiring people with these qualities is an option—one that more and more employers are taking into consideration, particularly for graduate trainees—provided we can assess them effectively. Flatter organizational structures have made it necessary to hire senior managers primarily on the basis of their track record rather than necessarily their interpersonal skills. This has resulted in the need for current "home grown" senior managers to become more people-oriented as well as business-oriented (Long *et al.*, 2013).

When organizations implement competency-based performance management systems, the requirement to quantify competencies becomes more apparent. Unsurprisingly, more than 50% of survey respondents said that their organizations evaluated job performance based on competencies. The consistency of evaluation and the clarity of the behavioral indications received the greatest user complaints in our survey, especially from respondents who had not received any training in the methodology. Can we correctly and consistently measure individual competencies given all the issues with assessment? (Loomba and Nakashima, 2012).

11.5 Conclusions

Any system support plan has always included regular maintenance as a key component. The majority of us have been doing recommended maintenance tasks without much consideration for the real state of the system or the remaining useful life of the parts that have been removed. We have been doing it because the maintenance manual instructs us to do it on a schedule or because we want to prevent unexpected breakdowns. This has resulted in enormous time and financial waste. We have focused the reader's attention in this work on CBM principles, which have been shown to be more effective than conventional maintenance methods in a number of ways. The concise survey of the literature shows how actively researchers are working to develop the technological capabilities that support CBM.

The process of creating a list of skills should involve a variety of professional social work organizations and provide time for input from all social workers. Making the assessment of capabilities meaningful and useful in condition-based maintenance and residual life prediction depends on carefully and completely establishing the list of basic competencies at both the foundational and the advanced or specialized levels. This is a noble and significant activity. The interaction between CBM and RLP necessitates a comprehensive evaluation of professional competency. Analyzing competencies and the best methods for assessing them can help work efforts to ensure improved outcomes. Strong professional organizations must clearly define competencies and develop reliable means to evaluate them. To serve and safeguard the services, as well as the profession, this is a crucial endeavor.

References

Ahmad, R. and Kamaruddin, S., An overview of time-based and condition-based maintenance in industrial application. *Comput. Ind. Eng.*, 63, 1, 135–149, 2012, doi: 10.1016/j.cie.2012.02.002.

Banerjee, P., Resource dependence and core competence: Insights from Indian software firms. *Technovation*, 23, 3, 251–263, 2003, doi: 10.1016/S0166-4972(01)00120-1.

Banjevic, D., *et al.*, A control-limit policy and software for condition-based maintenance optimization. *Infor*, 39, 1, 32–50, 2001, doi: 10.1080/03155986.2001.11732424.

Bengtsson, M., Condition Based Maintenance Systems: An Investigation of Technical Constituents and Organizational Aspects, Licentiate thesis Mälardalen University Eskilstuna, 2004, Available at: http://citeseerx.ist.psu.edu/viewdoc/download?doi=10.1.1.122.2614&rep=rep1&type=pdf%5Cnhttp://citeseerx.ist.psu.edu/viewdoc/download?doi=10.1.1.122.2614&rep=rep1&type=pdf.

Bertoncelj, A., Okvir menadžerskih nadležnosti: Studija konativne komponente. *Ekonomska Istrazivanja*, 23, 4, 91–101, 2010, doi: 10.1080/1331677X.2010.11517435.

Bhamra, R., Dani, S., Bhamra, T., Competence understanding and use in SMEs: A UK manufacturing perspective. *Int. J. Prod. Res.*, 49, 10, 2729–2743, 2011, doi: 10.1080/00207541003738873.

Boikov, A.V., The prediction of the residual life of electromechanical equipment based on the artificial neural network, *IOP Conf. Ser.: Earth Environ. Sci.*, 8, 203–222, 2017.

Chen, D. and Trivedi, K.S., Closed-form analytical results for condition-based maintenance. *Reliab. Eng. Syst. Saf.*, 76, 1, 43–51, 2002, doi: 10.1016/S0951-8320(01)00141-7.

Dongshan, Z., Residual strength calculation & residual life prediction of general corrosion pipeline. *Procedia Eng.*, 94, 52–57, 2014, doi: 10.1016/j.proeng.2013.11.045.

Elena Dragomir, O., *et al.*, Review of prognostic problem in condition-based maintenance. *IEEE Control Systems Society European Control Conference*, 1587–1592, 2009, Available at: https://hal.archives-ouvertes.fr/hal-00418761.

Ellis, B.A., Condition Based Maintenance, *Websiste?Magazine?. Jethroproject Group*, 24, 1–5, 2008.

Fouad, N.A., *et al.*, Competency Benchmarks: A Model for Understanding and Measuring Competence in Professional Psychology Across Training Levels. *Training Educ. Prof. Psychol.*, 3, 4, Suppl. 1, 19–30, 2009, doi: 10.1037/a0015832.

Gebraeel, N., *et al.*, Residual life predictions from vibration-based degradation signals: A neural network approach. *IEEE Trans. Ind. Electron.*, 51, 3, 694–700, 2004, doi: 10.1109/TIE.2004.824875.

Guo, B., Life cycle view of firm's core competence: Chinese manufacturing industry as a case. *Proceedings of the 2000 IEEE Engineering Management Society, EMS 2000*, pp. 363–367, 2000, doi: 10.1109/EMS.2000.872531.

Al Taee, H. and Jihad Alwaely, D.F., Examining the Relationships Among Human. *Global J. Manage. Bus. Res.*, 12, 23, 2012.

Hao, L., *et al.*, Residual life prediction of multistage manufacturing processes with interaction between tool wear and product quality degradation. *IEEE Trans. Autom. Sci. Eng.*, 14, 2, 1211–1224, 2017, doi: 10.1109/TASE.2015.2513208.

Herzog, M.A., Marwala, T., Heyns, P.S., Machine and component residual life estimation through the application of neural networks. *Reliab. Eng. Syst. Saf.*, 94, 2, 479–489, 2009, doi: 10.1016/j.ress.2008.05.008.

Huang, R., *et al.*, Residual life predictions for ball bearings based on self-organizing map and back propagation neural network methods. *Mech. Syst. Signal Process.*, 21, 1, 193–207, 2007, doi: 10.1016/j.ymssp.2005.11.008.

Hulla, M., *et al.*, A case study based digitalization training for learning factories. *Procedia Manuf.*, 31, 169–174, 2019, doi: 10.1016/j.promfg.2019.03.027.

Ilchenko, B.V., *et al.*, Residual life prediction of power steam turbine disk with fixed operating time, in: *Proceedings of 19th European Conference on Fracture*, ECF19. Citeseer, pp. 26–31, 2012.

Jain, V.K., Competency Mapping in Indian Industries-A Case Study. *Int. J. Emerg. Res. Manage. Technol.*, 2, 10, 9–21, 2013.

Jardine, A.K.S., Lin, D., Banjevic, D., A review on machinery diagnostics and prognostics implementing condition-based maintenance. *Mech. Syst. Signal Process.*, 20, 7, 1483–1510, 2006, doi: 10.1016/j.ymssp.2005.09.012.

Jaske, C.E., Predicting the Residual Life of Plant Equipment - Why Worry? *Am. Soc. Mech. Eng. Press. Vessels Pip. Div. (Publication) PVP*, 98–1, August, 3–8, 1985.

de Jonge, B., Teunter, R., Tinga, T., The influence of practical factors on the benefits of condition-based maintenance over time-based maintenance. *Reliab. Eng. Syst. Saf.*, 158, 21–30, 2017, doi: 10.1016/j.ress.2016.10.002.

Kaane, S., Marketing reference and information services in libraries: a staff competencies framework. *72nd IFLA General Conference*, Seoul, Korea, March, 2006, Available at: http://citeseerx.ist.psu.edu/viewdoc/download?doi=10.1.1.150.6905&rep=rep1&type=pdf.

Liu, F., *et al.*, Residual life prediction for ball bearings based on joint approximate diagonalization of eigen matrices and extreme learning machine. *Proc. Inst. Mech. Eng., Part C: J. Mech. Eng. Sci.*, 231, 9, 1699–1711, 2017, doi: 10.1177/0954406215621585.

Long, C.S., Wan Ismail, W.K., Amin, S.M., The role of change agent as mediator in the relationship between HR competencies and organizational performance. *Int. J. Hum. Resour. Manage.*, 24, 10, 2019–2033, 2013, doi: 10.1080/09585192.2012.725080.

Loomba, A.P.S. and Nakashima, K., Enhancing value in reverse supply chains by sorting before product recovery. *Prod. Plann. Control*, 23, 2–3, 205–215, 2012, doi: 10.1080/09537287.2011.591652.

Loué, C. and Baronet, J., Toward a new entrepreneurial skills and competencies framework: A qualitative and quantitative study. *Int. J. Entrepreneurship Small Bus.*, 17, 4, 455–477, 2012, doi: 10.1504/IJESB.2012.050164.

Manna, R., Singh, A., Sharma, P., Does Training Need Analysis Help to Minimize Competency Gap: An Investigation. *AJTD ADMAA Amity J. Train. Dev.*, 1, 1, 109–131, 2016.

Miranda, J., *et al.*, The core components of education 4.0 in higher education: Three case studies in engineering education. *Comput. Electr. Eng.*, 93, September 2020, 2021, doi: 10.1016/j.compeleceng.2021.107278.

Mohanty, J.R., Verma, B.B., Ray, P.K., Prediction of fatigue crack growth and residual life using an exponential model (constant amplitude loading). *Int. J. Fatigue*, 31, 3, 418–424, 2009, doi: 10.1016/j.ijfatigue.2008.07.015.

Momme, J., Framework for outsourcing manufacturing: Strategic and operational implications. *Comput. Ind.*, 49, 1, 59–75, 2002, doi: 10.1016/S0166-3615(02)00059-3.

Nurach, P., Thawesaengskulthai, D., Chandrachai, A., Developing an Organization Competencies Framework for SME(s) in Thailand. *Chin. Bus. Rev.*, 11, 02, 233–241, 2012, doi: 10.17265/1537-1506/2012.02.010.

Pawar, P.M. and Ganguli, R., Fuzzy-logic-based health monitoring and residual-life prediction for composite helicopter rotor. *J. Aircr.*, 44, 3, 981–995, 2007, doi: 10.2514/1.26495.

Prajapati, A., Bechtel, J., Ganesan, S., Condition based maintenance: A survey. *J. Qual. Maint. Eng.*, 18, 4, 384–400, 2012, doi: 10.1108/13552511211281552.

Ray, A.K., *et al.*, Residual life prediction of service exposed main steam pipe of boilers in a thermal power plant. *Eng. Fail. Anal.*, 7, 5, 359–376, 2000, doi: 10.1016/S1350-6307(99)00024-2.

Sehr, E.M., Competency development in technology management for enhancing competitiveness. *AIMA J. Manage. Res.*, 11, 4, 1–17, 2005.

Shah, C.R., Competency Mapping and its impact on Organisation Effectiveness with special reference to Sales Staff in Pharmaceutical Industry of Ahmedabad Region, 2016, Available at: http://www.gtu.ac.in/uploads/Synopsis - 119997392047 - Chaitali Shah.pdf.

Shaheen, M., *et al.*, A competency framework for contractual workers of manufacturing sector. *Ind. Commer. Training*, 51, 3, 152–164, 2019, doi: 10.1108/ICT-10-2018-0080.

Sigh, R.K., Garg, S.K., Deshmukh, S.G., Challeges and strategies for competitiveness of SMEs: A case study in the Indian context. *Int. J. Serv. Oper. Manage.*, 4, 2, 181–200, 2008, doi: 10.1504/IJSOM.2008.016610.

Singh, C.D. and Khamba, J.S., A case study of a two wheeler manufacturing unit on manufacturing competency and strategic success. *Int. J. Eng. Res. Afr.*, 19, 156–170, 2016, doi: 10.4028/www.scientific.net/JERA.19.156.

Tao, C., *et al.*, Characterization of fatigue damages in composite laminates using Lamb wave velocity and prediction of residual life. *Compos. Struct.*, 166, November 2020, 219–228, 2017, doi: 10.1016/j.compstruct.2017.01.034.

Vinet, L. and Zhedanov, A., A "missing" family of classical orthogonal polynomials. *J. Phys. A: Math. Theor.*, 44, 8, 1–15, 2011, doi: 10.1088/1751-8113/44/8/085201.

Wang, W., *et al.*, A model for residual life prediction based on Brownian motion with an adaptive drift. *Microelectron. Reliab.*, 51, 2, 285–293, 2011, doi: 10.1016/j.microrel.2010.09.013.

Yam, R.C.M., *et al.*, Intelligent predictive decision support system for condition-based maintenance. *Int. J. Adv. Manuf. Technol.*, 17, 5, 383–391, 2001, doi: 10.1007/s001700170173.

Yamani, H., The Study of Digital Transformation Skills and Competencies Framework at Umm Alqura University. *Int. J. Econ. Manag. Eng.*, 475–479, 2021, Available at: https://www.researchgate.net/publication/348805463.

Meng, G., A generalized similarity measure for similarity-based residual life prediction. *Proc. Inst. Mech. Eng., Part E: J. Process Mech. Eng.*, 225, 3, 151–160, 2011, doi: 10.1177/0954408911399832.

Zhou, R., Serban, N., Gebraeel, N., Degradation-based residual life prediction under different environments. *Ann. Appl. Stat.*, 8, 3, 1671–1689, 2014, doi: 10.1214/14-AOAS749.

12

Sustainability Issues in CBM and RLP: Case Studies

Simranjit Singh Sidhu[1]* and Gurpreet Singh Sidhu[2]

[1]MED, Punjabi University Patiala, Punjab, India
[2]Punjab Institute of Technology, Rajpura, Punjab, India

Abstract

Two case studies have been conducted to ascertain the step-by-step implementation of maintenance practices adopted by organizations toward achieving their industrial objectives. For this, support was sought from various reputed northern Indian manufacturing SMEs. For conducting the case study, the detailed research objectives, methodology, and scope of work were shared and discussed with the concerned organizations' resource persons, followed by detailed clarification and personal interactions. In response, the following enterprises have given their consent and provided their required letter of support for conducting the case study:

- Marelli India Pvt. Ltd., (Plot No, 402, Sector 8 Rd, Sector 8, Manesar, Gurugram, Haryana) (medium enterprise) and
- Gurkirpa Industries (D139, Phase 7, Industrial Area, Sector 73, Sahibzada Ajit Singh Nagar, Punjab) Mohali (Punjab, India) (small-scale organization).

This chapter presents descriptive case studies conducted in leading SMEs that have implemented maintenance practices over a considerable time and have realized holistic performance enhancements as a result of implementing maintenance practices. The objective of the case studies was to investigate the status of maintenance practices implementation and compare benefits attained through the implementation of this approach in terms of improvements in manufacturing performance parameters. Further, this study focuses on the techniques/

**Corresponding author*: Simranjit_me@pbi.ac.in

Chandan Deep Singh, Davinder Singh, Kanwal Jit Singh and Harleen Kaur (eds.) Condition-Based Maintenance and Residual Life Prediction, (233–272) © 2025 Scrivener Publishing LLC

methodologies adopted for implementing maintenance practices and finally, the benefits achieved in compliance.

Manufacturing technologies have undergone gradual but revolutionary changes worldwide, and developing countries like India are also responding to these changes. India has surpassed Japan to become the world's fourth largest vehicle market, with sales jumping 8.3% year on year to 3.99 million units (source: India Brand Equity Foundation). From April 2000 to June 2019, the automobile industry attracted US $22.35 billion in foreign direct investment (FDI) (Department for Promotion of Industry and Internal Trade, Government of India). In 2019, India exported 676,193 passenger automobiles, and the exports accounted for 26% of the market by 2021. Furthermore, a cost-effective manufacturing base keeps costs down by 10%–25% compared to Europe and the United States. In 2018–2019, the auto-components business contributed 2.3% to India's GDP, 25% of its manufacturing GDP, and employed 50 lakh people. The automobile component exports from India are estimated to reach US $80 billion by 2026, according to the Automobile Component Manufacturers Association (ACMA).

***Keywords*:** Case studies, maintenance practices, sustainability, MTTR, MTBF, training, skill development, and SWOT analyses

12.1 Medium Industry Case Study

The automotive component Marelli India Pvt. Ltd. selected for the study falls under the medium-scale manufacturing industry and is one of the country's largest suppliers of quality auto parts. A medium enterprise is defined as an investment in plant and machinery or equipment of less than 50 crore rupees and a turnover of less than 250 crore rupees (Ministry of Micro, Small and Medium Enterprises, Government of India, 1st June, 2020).

The company offers farm equipment, automotive, automotive lighting, electronic systems, powertrain, suspension systems, shock absorbers, SDC (suspension; shock absorber; synaptic damping control), exhaust systems, automotive lighting, cabin comfort, electric powertrain, electronic systems, and ignition. Swaraj group, Ashok Leyland, Mahindra and Mahindra, Tata, Maruti Suzuki, and Honda are some of the company's valued customers.

The company's customers frequently audit the company's performance on some key performance indicators. Drawing the importance and benefits of key maintenance practices in reliability, and to increase the overall equipment effectiveness (OEE), the company decided to implement strategic maintenance practices in a phase-wise manner to resolve this issue. A team was formulated under the leadership of top management, comprising of key persons from various departments, as depicted in Figure 12.1.

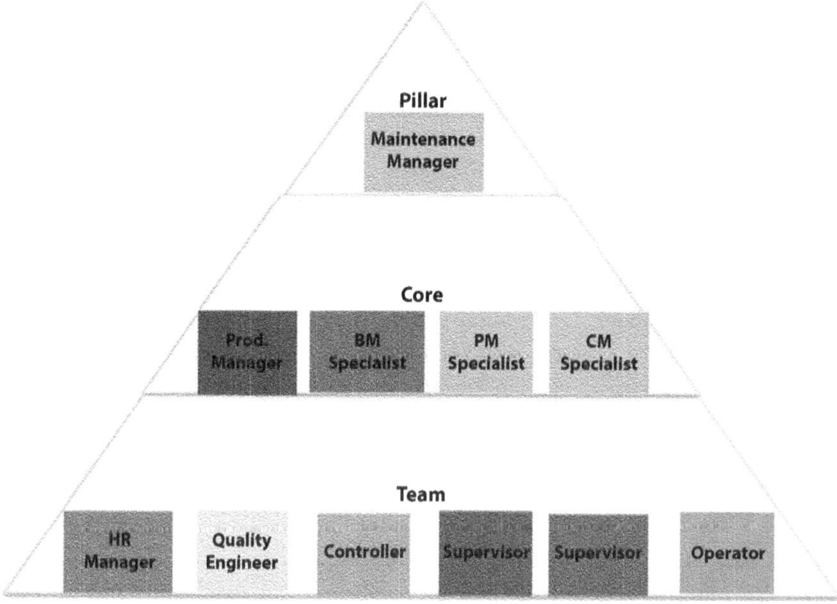

Figure 12.1 Maintenance practices team.

The main reasons behind overhauling maintenance practices were increased productivity, decreased breakdowns, and increased equipment availability. A number of brainstorming sessions were held to find the solution for increasing productivity, decreasing breakdowns, and increasing equipment availability. Table 12.1 reveals the phase-wise maintenance improvement initiatives envisaged by the manufacturing organization, along with the measurable outcomes from the successful implementation of this initiative.

12.2 Objectives of Implementing Maintenance Improvement Initiatives

- Maximize machine availability
- Improve plant efficiency (productivity)
- Eliminate losses due to lack of maintenance
- Increase mean time between failures (MTBF)
- Decrease mean time to repair (MTTR)

Table 12.1 Phase-wise maintenance improvement initiatives.

Phase	Name of phase	Implemented initiatives	Output
1.	Transition phase	General maintenance issues, breakdown maintenance (BM) issues, and preventive maintenance (PM) issues	Production and quality; safety and morale
2.	Intermediate phase	Predictive maintenance (Pd.M) issues, corrective maintenance (CM) issues, and computer maintenance management system (CMMS) issues	Cost, flexibility, and delivery; OEE
3.	Maturity phase	Maintenance scheduling issues and total productive maintenance issues	Availability and reliability; reduction in operation losses overhead equipment expenditure; manpower Losses; and business performance

12.3 Need for Maintenance

Manufacturing maintenance is an important tool that assumes a basic part in an organization's capacity to contend on the premise of cost, quality, and execution of delivery. Maintenance aims to contribute toward a company's profit, as maintenance is the most required function that the organization should opt to attain a pioneer position among its competitors (Dixit and Raj, 2019). Maintenance has become a major budgetary variable for any company, with maintenance costs ranging from 5% to 20% of total production costs (Ahuja and Khamba, 2008). Maintenance has a critical role in manufacturing industries because its impact on other working areas such as production, quality, production cost, working environment, and the amount of work-in-progress significantly impacts the company's overall sustainability endeavors.

Maintenance also aids in improving production quality, which is envisaged as a critical factor in determining an organization's success.

12.4 Phase-Wise Implementation of Maintenance Practices

12.4.1 Phase 1: Transition

In phase 1 of this study, general maintenance issues, breakdown maintenance (BM) issues, and preventive maintenance (PM) issues have been assessed.

General maintenance has established short, medium, and long-term goals, and has ensured that employees are aware of their responsibilities for achieving effective maintenance. The top management has also successfully defined and percolated the maintenance objectives up to grass-root level, to clearly define each employee's role. As a result, appropriate spares, supplies, and equipment and tooling required for maintenance have been made available on all workstations (Crocitto and Youssef, 2003; Kedar *et al.*, 2016). SMEs in India have implemented all the general maintenance issues in an appropriate manner (Baglee and Knowles, 2010).

Breakdown maintenance (BM) supports the methodology, where maintenance is done when the apparatus stops and causes a severe decay (Wireman, 1990). When a breakdown occurs, inventory parts are readily available, and the shutdowns are handled in an optimum time period. The breakdown maintenance methodology adopted by the organization is depicted in Figure 12.2.

The preventive maintenance initiatives involved operators working on minor PMs such as cleaning, lubricating, adjusting, and inspecting PM check sheets, thereby extensively implementing PM tasks for each piece of equipment. The countermeasures were effectively deployed under the PM program to achieve zero breakdowns (Bris *et al.*, 2003; Ahuja and Khamba, 2008). In the SME sector, well-planned checklists for maintenance and inspection have been used extensively to eliminate routine and chronic breakdowns by implementing PM programs.

During the implementation of Phase 1 of the case study, it was noticed that the machine cover was missing, as shown in Figure 12.3. The new cover inserted on the head guide of the machine facilitated enhanced safety of the workers. One more problem identified was that the motor cover was broken, as shown in Figure 12.4. The new cover was inserted on the motor.

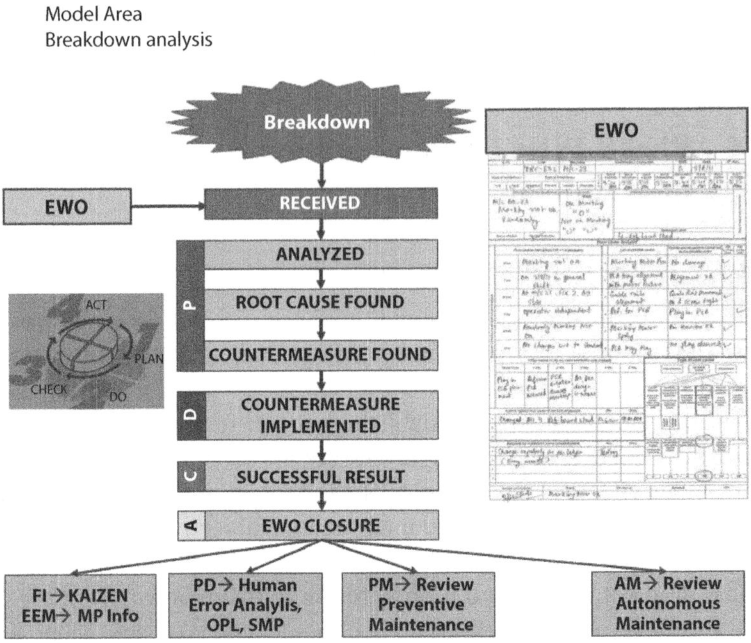

Figure 12.2 Breakdown analysis.

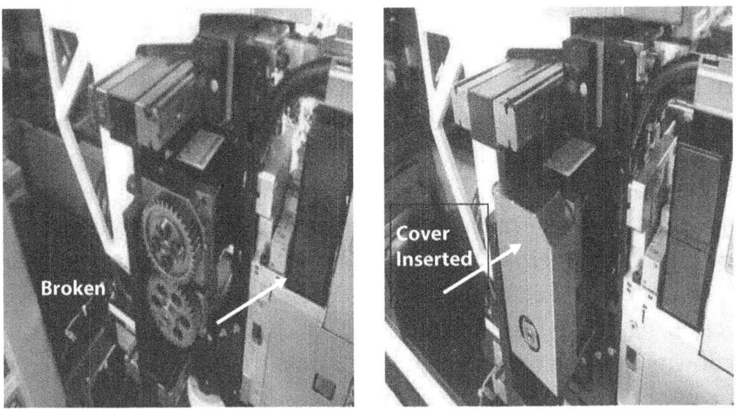

Figure 12.3 The cover was missing and a new one was inserted on the head guide of the machine.

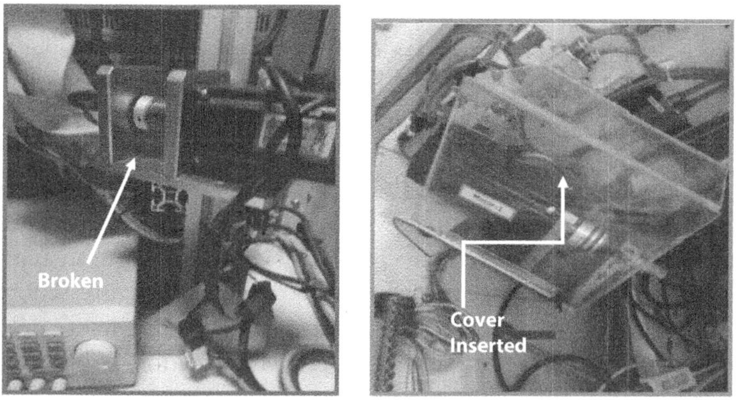

Figure 12.4 The motor cover was broken and a new one was inserted on the motor.

During a core team meeting, it was discovered that the previous order flow for a new machine with precise specifications is as depicted in Figure 12.5. The drawback of this process was that there was no direct connection between the PM and the supplier. To get over this limitation, a new flow was created that provided a direct relationship between the PM and the supplier. This way, a timely response is obtained from the supplier and innovative solutions are integrated on the machine being manufactured.

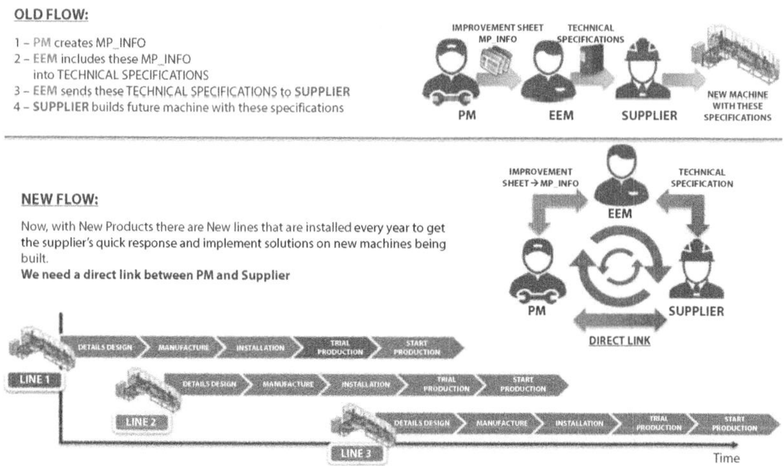

Figure 12.5 Flow of order of a new machine with a particular specification.

Furthermore, as illustrated in Figure 12.6, dust-free tool inventory and storage have been established. Following that, a spare part management program was implemented, in which spare parts were organized and made easily available whenever a breakdown occurs, as depicted in Figures 12.7 and 12.8, respectively.

Another problem hindering the maintenance function performance was inappropriate recordkeeping, which causes extended delays and high searching times. This problem was resolved by implementing an appropriate 5S housekeeping storage mechanism, as depicted in Figure 12.9.

Another key effort taken during the first phase of maintenance practices implementation was the organization's preventive action to boost equipment reliability. A mist filter was used to prevent moisture from entering the machine to extend the machine's life, as shown in Figure 12.10.

The organization has directed its efforts toward identifying crucial areas for improvement to increase the plant's dependability and productivity. MTTR and MTBF are two crucial areas for further development in this phase. The procedure for the evaluation of MTTR is shown in Figure 12.11.

Figure 12.6 Storage is modified.

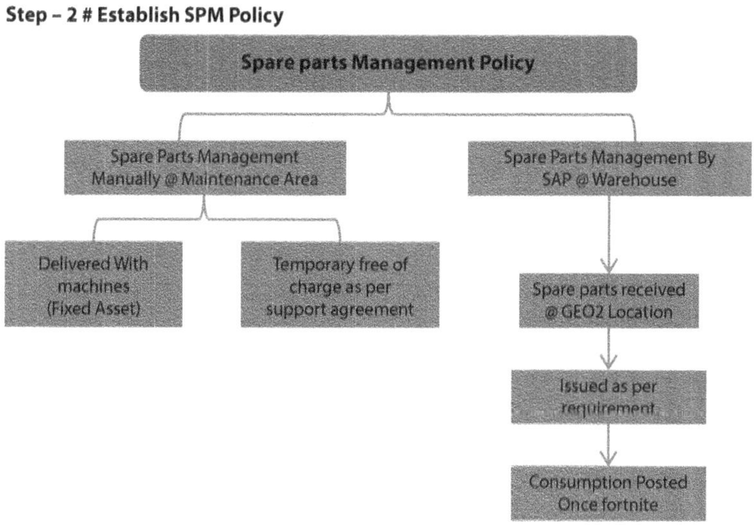

Figure 12.7 Spare parts management.

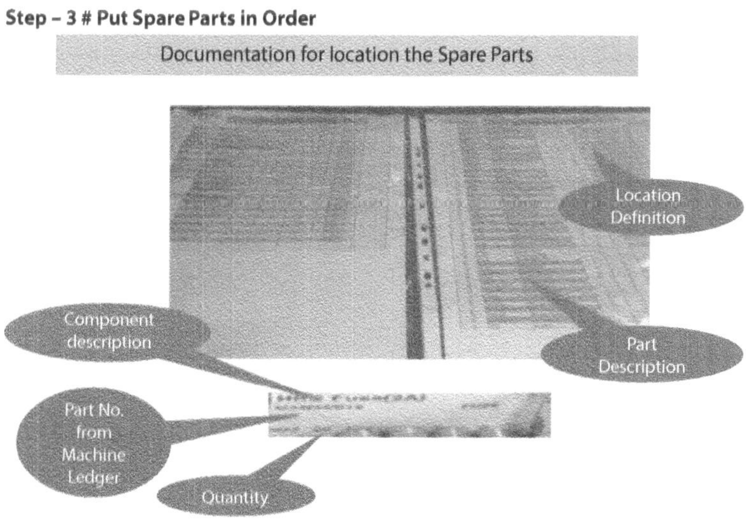

Figure 12.8 Spare parts in order.

Figure 12.9 Storage of manuals and drawings.

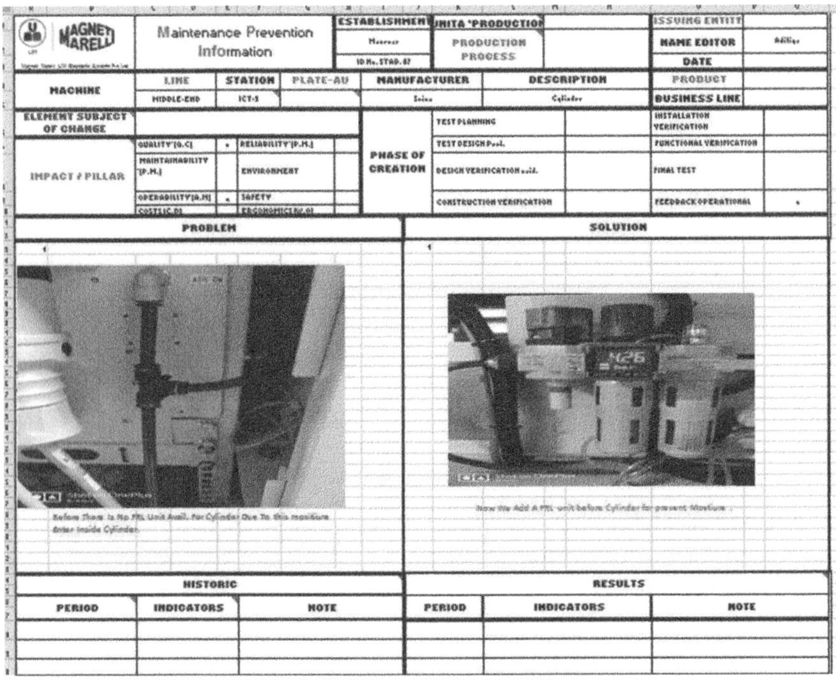
MAGNETI MARELLI
Maintenance Prevention Information
ESTABLISHMENT
ID No. STAD. 87
UNITA 'PRODUCTION
PRODUCTION PROCESS
ISSUING ENTITY
NAME EDITOR
DATE
MACHINE
LINE
MIDDLE-END
STATION
ICT-1
PLATE-AU
MANUFACTURER
DESCRIPTION
Cylinder
PRODUCT
BUSINESS LINE
ELEMENT SUBJECT OF CHANGE
IMPACT / PILLAR
QUALITY [Q.C]
RELIABILITY [P.M.]
MAINTAINABILITY [P.M.]
ENVIRONMENT
OPERABILITY [A.M]
SAFETY
COSTS [C.D]
ERGONOMICS
PHASE OF CREATION
TEST PLANNING
TEST DESIGN
DESIGN VERIFICATION
CONSTRUCTION VERIFICATION
INSTALLATION VERIFICATION
FUNCTIONAL VERIFICATION
FINAL TEST
FEEDBACK OPERATIONAL
PROBLEM
SOLUTION
HISTORIC
PERIOD
INDICATORS
NOTE
RESULTS
PERIOD
INDICATORS
NOTE

Figure 12.10 Use of mist filter before machine air pressure to prevent moisture.

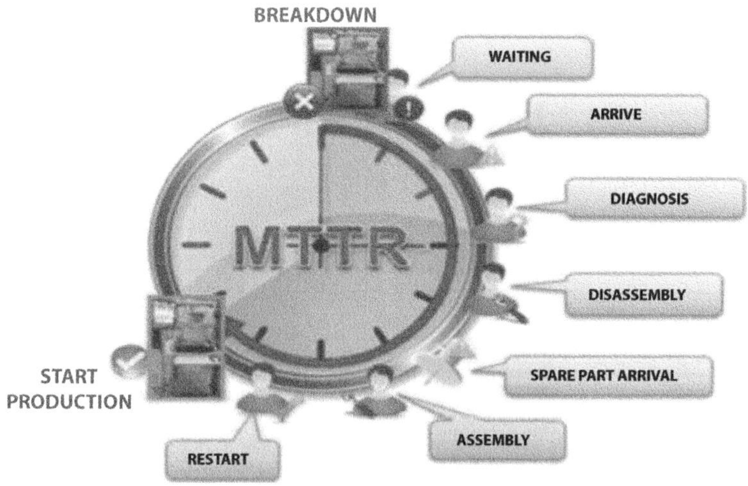

Figure 12.11 Procedure of the mean time to repair.

Mean time to repair (MTTR) is a maintenance metric that assesses how long it takes to troubleshoot and fix faulty equipment, on average. MTTR is 148 minutes, as shown in Figure 12.12.

The average MTTR has been reduced to 128 minutes in 2021, a 19% reduction, through the appropriate implementation of suitable maintenance methods. The reduced MTTR contributes to increased company productivity as shown in Figure 12.13.

The mean time between failures (MTBF) is the average time between system failures. MTBF is a vital maintenance indicator for assessing performance, safety, and equipment design.

12.4.2 Phase 2: Intermediate

As a result of the improvements achieved in phase 1, the company identified the critical areas that needed improvement in phase 2. The company implemented predictive maintenance (Pd.M) issues, corrective maintenance (CM) issues and computer maintenance management system (CMMS) issues to decrease the cost of product and increase OEE, flexibility, and delivery.

During corrective maintenance, it was observed that there was a play in the fixture when the fixture tray is moved. During machine operations, the fixture components: the motor and pallet stud are prone to friction. The design of a new fixture rail was developed. Fixture locking was activated

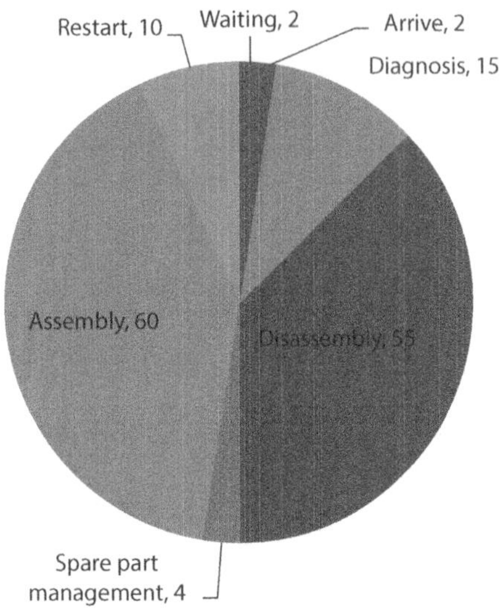

Figure 12.12 Mean time to repair.

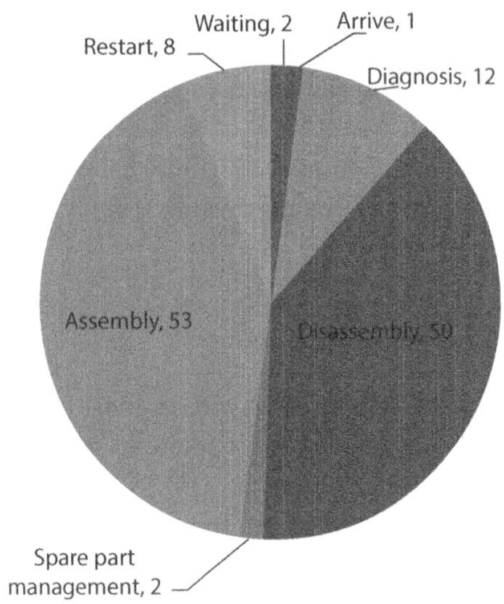

Figure 12.13 Reduced mean time to repair.

by a limit switch that detects the fixture close position and engages locking cylinders, as shown in Figure 12.14. An improved locking system guarantees solid operation during tests and steady movement of the fixture tray when it is opened and closed. For locking, the guide stoppers were connected with the fixture rail. The motor and pallet stud life was enhanced from 6 to 12 months after the design changes.

As shown in Figures 12.15–12.17, the standard maintenance procedure (SMP) was employed for pin replacement. A fixture test for pin replacement was performed to detect the damaged pin. Then, the fixture was locked and the machine was kept in manual mode. After that, the knob was turned. The fixture from the machine was also removed and it was kept on the cart. The final step was to inspect the condition of the pin, and if necessary, it was replaced with a new one, which was occasionally encountered. The pin was caught inside the receptacle. The lack of PM was the root of the problem. Pin change frequency was not specified. The frequency of changing the pin was defined as a preventive measure.

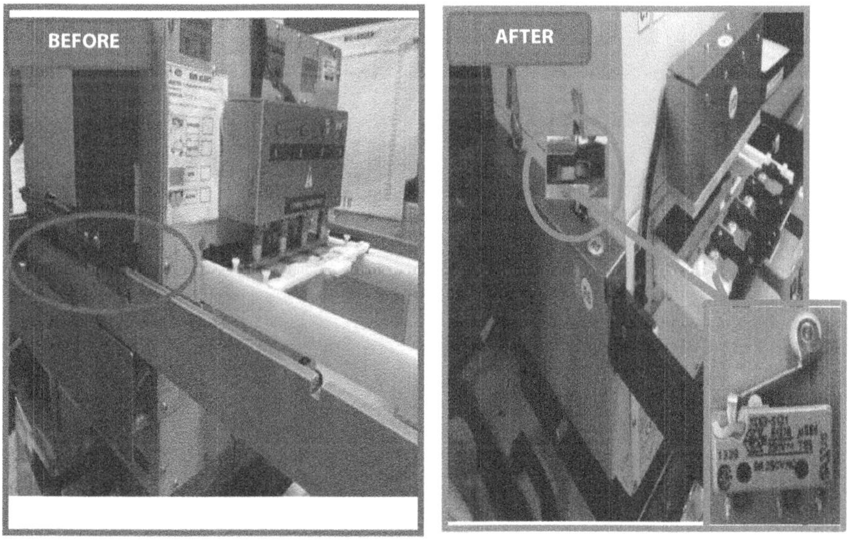

Figure 12.14 New design of fixture rail to remove the play in the fixture during the movement of the fixture tray.

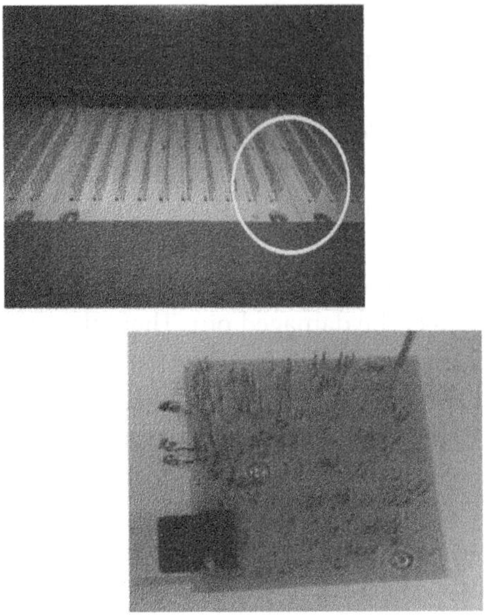

Figure 12.15 Replacement of defective pin.

Figure 12.16 Middle end ICT-1 machine.

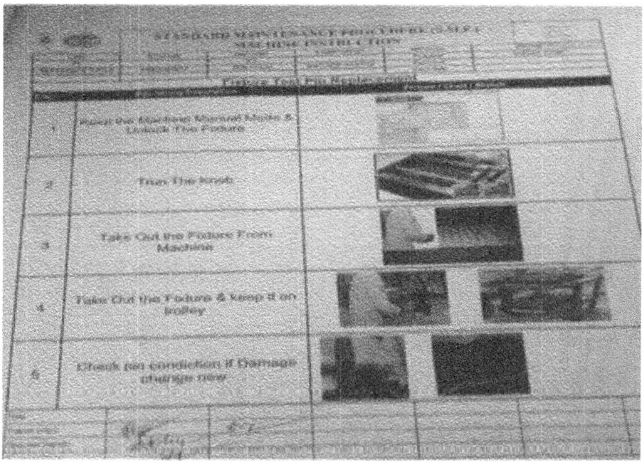

Figure 12.17 Standard maintenance procedure (SMP).

As shown in Figure 12.18, the number of breakdowns has decreased in phase 2 due to the deployment of maintenance methods. As seen in Figure 12.18, preventive and corrective maintenance helped increase the MTBF.

In phase 2, OEE has increased because of better maintenance methods, which has resulted in lower costs and improved product flexibility and delivery, as shown in Figure 12.19.

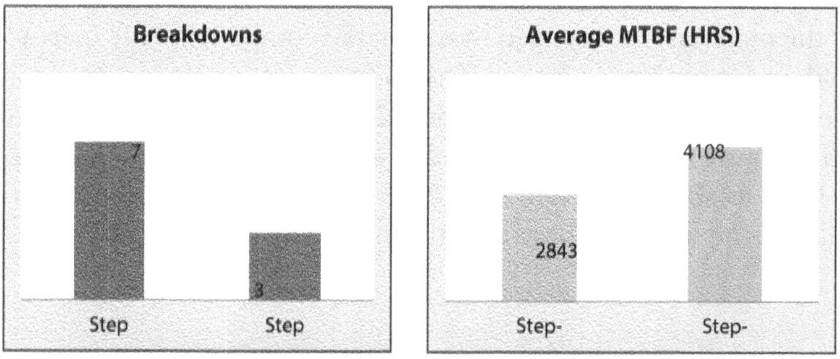

Figure 12.18 Decrease in breakdowns and increase in average MTBF.

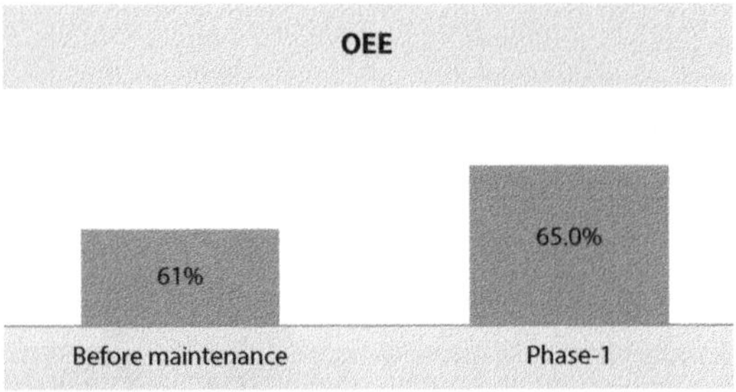

Figure 12.19 Overall equipment efficiency.

12.4.3 Phase 3: Maturity

Maintenance practices were continued in the third phase to reap the most benefits. Due to the maintenance schedule and TPM, OEE has improved. It reduced the average amount of time spent idle; the availability and dependability of the system have also been increased. Prior to the implementation of maintenance practices, overhead equipment expenditures, workforce losses, business performance, and operational costs all suffered due to reduced operations.

The focus was on extending the component's average life as shown in Figures 12.20 and 12.21. After a brainstorming session, modifying the component's substance was agreed. The substance of the reference board stud has changed from Derlin to strong Nylon. Previously, they had reference material board stud made of Derlin and had a life of around 1 month. After the material was changed to nylon, the component's life increased to more than 6 months.

Previously, the PM activity for the soldering head was conducted every month because solder wire flux residue accumulated at the soldering head area. Now, as shown in Figures 12.22 and 12.23, with the fume extractor installed at the soldering head mouth, smoke will go directly into the fume

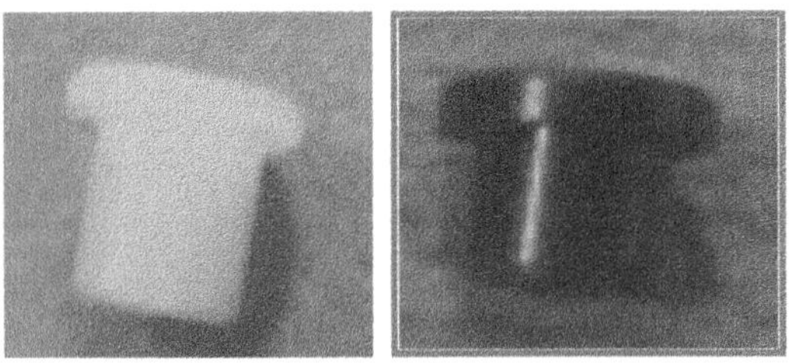

Figure 12.20 Reference board stud of two different materials.

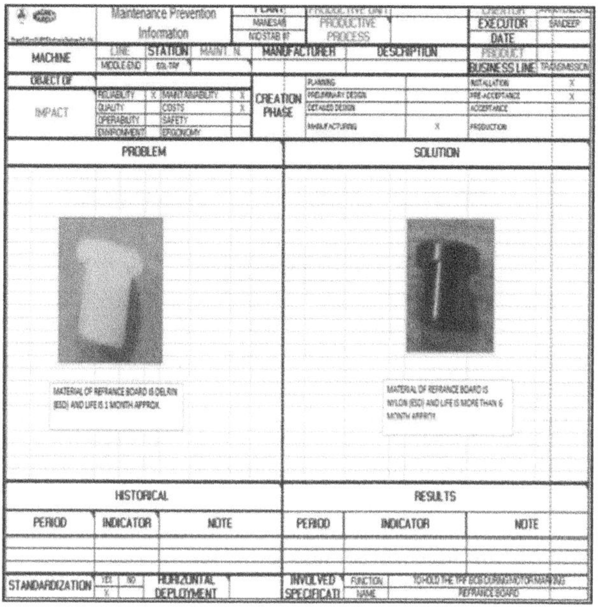
Maintenance Prevention Information
MANESAR
PRODUCTIVE PROCESS
EXECUTOR SANDEEP
DATE
MACHINE
LINE
STATION
MANUFACTURER
DESCRIPTION
PRODUCT
BUSINESS LINE
OBJECT OF
IMPACT
RELIABILITY X
MAINTAINABILITY X
QUALITY
COSTS X
OPERABILITY
SAFETY
ENVIRONMENT
ERGONOMY
CREATION PHASE
PLANNING
PRELIMINARY DESIGN
DETAILED DESIGN
MANUFACTURING X
INSTALLATION X
PRE-ACCEPTANCE X
ACCEPTANCE
PRODUCTION
PROBLEM
SOLUTION
MATERIAL OF REFRANCE BOARD IS DELRIN (ESD) AND LIFE IS 1 MONTH APPROX.
MATERIAL OF REFRANCE BOARD IS NYLON (ESD) AND LIFE IS MORE THAN 6 MONTH APPROX.
HISTORICAL
RESULTS
PERIOD
INDICATOR
NOTE
PERIOD
INDICATOR
NOTE
STANDARDIZATION
YES
NO
X
HORIZONTAL DEPLOYMENT
INVOLVED SPECIFICATI
FUNCTION
NAME
REFRANCE BOARD

Figure 12.21 Reference board stud's sheet.

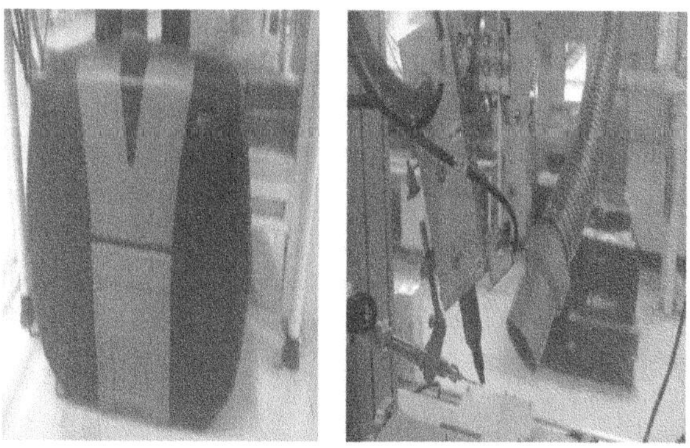

Figure 12.22 Fume extractor.

extractor, and very little residue will be generated, so they have increased the cleaning activity in the PM plan to 3 months. The expense of cleaning the head is lowered, and the installation of a fume extractor saves 44,500 rupees each year, as depicted in Table 12.2.

Figure 12.23 Fume extractor in the machine.

Table 12.2 Cost reduction by installation of fume extractor in the machine.

Maximum annual maintenance contract (AMC) cost (INR/year)	Before	After
Cost of cleaning head	124,300	83,700
Engineer's cost	8,400	4,500
Total cost of AMC	132,700	88,200
Amount saved	--	44,500

Further, to increase the machine's life and decrease breakdowns, the lubricant was replaced as shown in Figure 12.24. Biral T & D replaced WD40 for lubrication, which they were using previously and which resulted to a long time to clean and lubricate, causing poor head and syringe outcomes. With this change, the life of vacuum filters was extended with the use of

Figure 12.24 The lubricant oil was changed.

a brush feeder. The machine's operating costs were also reduced. The cost saving of changing lubrication oil is depicted in Table 12.3.

Further, when a breakdown does not occur, MTTR will also reduce to zero, as depicted in Figure 12.25. The preventive procedures applied in the third phase of maintenance practices also help increase the MTBF and OEE, as illustrated in Figure 12.26.

Table 12.3 Cost reduction by installation of fume extractor in the machine.

AMC cost (INR/Year)	Before	After
Cost of head cleaning	139,500	115,600
Engineer's cost	2,800	11,200
Total cost	142,300	126,800
Amount saved (in rupees)		15,500

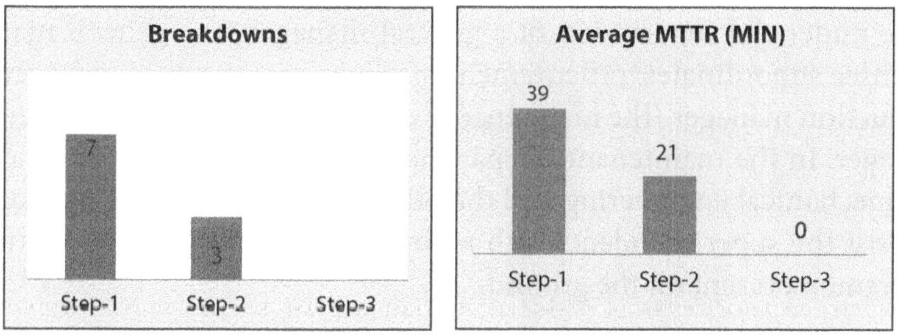

Figure 12.25 Decrease in breakdowns and increase in average MTBF.

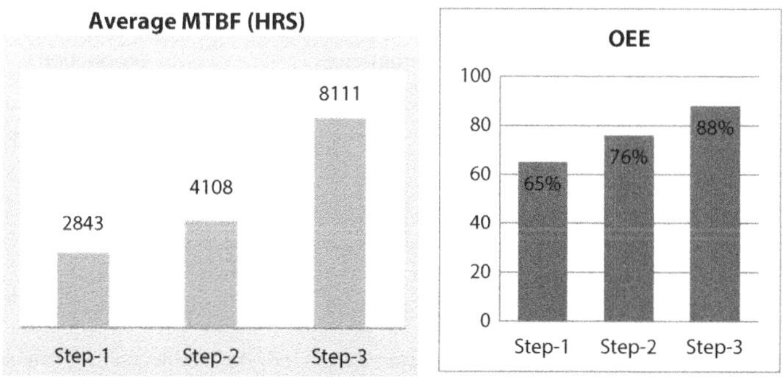

Figure 12.26 Increase in average MTBF and increase in OEE.

12.5 Small Industry Case Study

The case company "Gurkirpa Industries", is a small-scale enterprise based in northern India and established in 1990. Listed as a successful manufacturer, exporter, and supplier of a comprehensive assortment of spices grinding machine, chemical grinding machine, Besan grinding impact pulverizer machine, mixer machine, screw worm belt conveyor machine, Dalia making machine, cattle feed, and poultry feed grinding machine, rice husk bran grinding pulverizer machine, etc.

The company's product range is known for durability and performance under all conditions. These inclusive ranges of products cater to many small- and medium-scale industries in India with a number of specifications. The products are manufactured by the company using the best grade components and sophisticated technology in complete accordance with the set Confederation of Indian industry (CII) standards. The products are highly acknowledged for diverse attributes such as supreme quality, enhanced durability, corrosion resistance, low maintenance, and robust construction. The hierarchy of the maintenance organization is given in Figure 12.27.

There are a number of departments in the industry. All departments come under the supervision of a general manager. The general manager has three subordinates: procurement manager, maintenance manager, and production manager. The maintenance department is led by a maintenance manager. In the maintenance department, there are two superintendents, one mechanical engineering and the other electrical engineering. Further, to assist the superintendents with maintenance work, there are a number of foremen present on the ground.

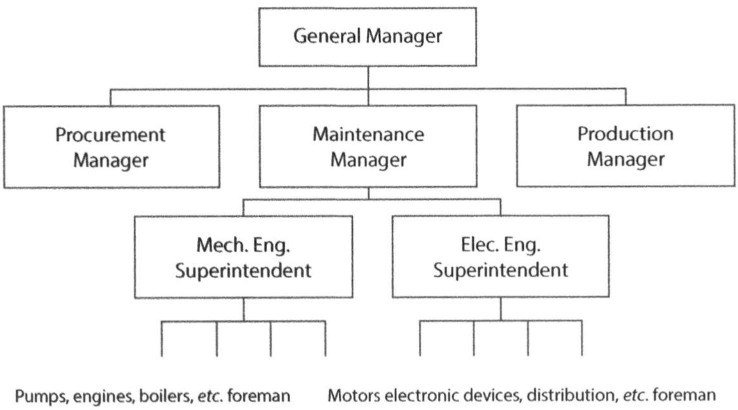

Figure 12.27 Hierarchy of the organization for maintenance.

Globalization has caused some recent changes in business originating from fluctuations in the market toward the increase in productivity and on time delivery of products. All these factors added pressure to the manufacturer to increase productivity and to deliver the product to the market on time. If this is not done, then the customer can shift loyalty toward other products. These circumstances forced the company to enhance its maintenance practices to increase the overall equipment effectiveness (OEE), which in turn will help retain loyal customers and help them in improving the firm's performance. The present study focuses on implementing various maintenance practices in the organization. For this, the management decided to review its existing maintenance processes and improve them in the organization, to increase productivity and reduce breakdown time to deliver timely products to customers.

12.6 Research Methodology

Case study description

A methodology was developed to analyze the existing system and develop future plans. A detailed step-by-step plan was prepared by researchers in discussion with the top management. Looking at the number of constraints faced by the organization, conducting a SWOT analysis (Kaczmarek and Małgorzata, 2016) was decided. Looking at the competitive market, some improvement areas still prevailed for future growth. A SWOT (strengths, weaknesses, opportunities, and threats) analysis of the enterprise was carried out, and then, based on the results, the adopted practices in the present is depicted in Figure 12.28. This investigation led to certain areas where the enhancement can be done in quite a straightforward manner (Table 12.4). From there, the company's issues were identified and examined. With the help of the workforce, majority of them were eradicated (Gupta and Mishra, 2016).

The methodology adopted by the case study company to affect performance enhancement through the adoption of strategic maintenance initiatives has been depicted in Figure 12.29. The step-by-step methodology included determining the critical areas that influence production and breakdowns; examining the current situation to identify what may be done to make it better; setting up appropriate maintenance practices; comparing and contrasting old and new circumstances if there is a positive change; and maintaining the new scenario; if the situation does not improve, appropriate maintenance processes are implemented.

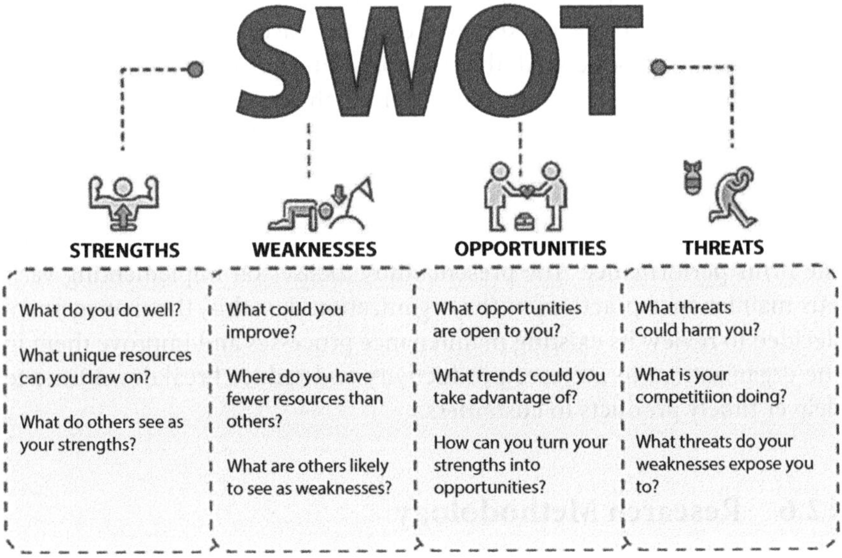

Figure 12.28 SWOT analysis (https://www.feedough.com/swot-analysis/).

Table 12.4 SWOT analysis.

Strengths	Weaknesses	Opportunities	Threats
Vision and mission	Frequent breakdowns	Market survey	Presence of competitors
Continuous upgrading	Low productivity	Competitive cost	Existence of Chinese products
Infrastructure	Less overall equipment efficiency	Delivery of products on time	Technology upgrade with time
Market image	Less skilled professionals		
Focus on improvement areas	Absence of maintenance training programs		
Multitasking employees	Upgrading is not timely		

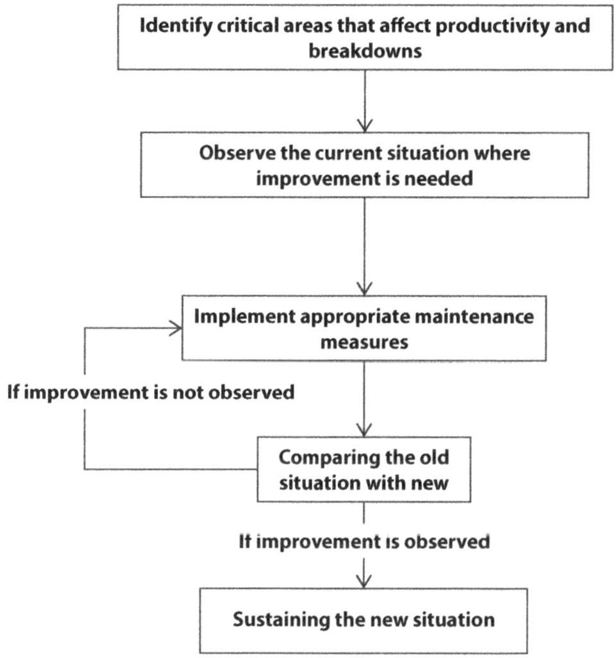

Figure 12.29 Strategic maintenance initiatives adopted by the organization.

12.7 Steps to Improve the Weaknesses Identified Through SWOT Analysis

Appropriate maintenance measures were implemented on the lathe machine.

As shown in Figure 12.30, the wear out of the gear in the lathe machine in the company led to machine breakdown, and work was halted. Earlier, there was no proper lubrication and no proper machine inspection. There was not any scheduling of preventive maintenance. There was not any check sheet for maintenance practices. Due to a lack of maintenance, there was gear teeth wear out as depicted in Figure 12.31. The company implemented maintenance practices to prevent sudden breakdowns of machines. Preventive maintenance in which oiling and lubrication are used was performed (Figure 12.32). It is essential to check the shaft alignment regularly. In preventive maintenance, gears are monitored regularly. The preventive maintenance checklist was prepared and holistically implemented in Table 12.5. After preventive maintenance was implemented, gears are changed after every 2 years.

Figure 12.30 Lathe machine halted due to gear breakdown.

Figure 12.31 Wear out of the gear teeth.

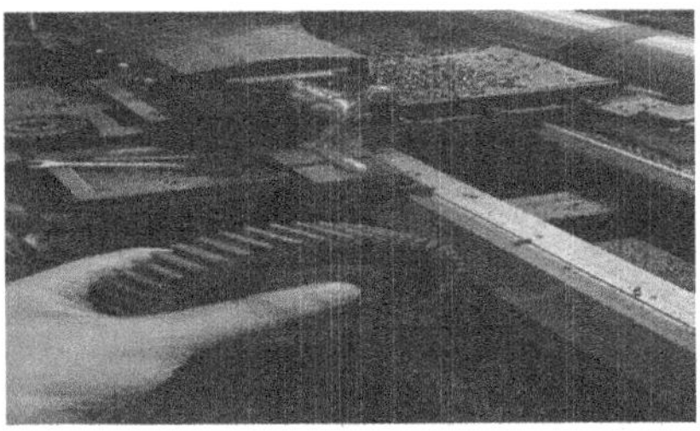

Figure 12.32 Gear replacement.

Further steps are taken to improve the reliability of the machine:

1. Cleaning has been done regularly.
2. Lubricating has been done regularly.
3. Adjusting and inspecting the machine have been done regularly.

Table 12.5 The preventive maintenance checklist.

Preventive Maintenance Check List (Date May 31/

Serial No.	Action	Check	Comments and reading	Inspected By
1.	Cleaning	ok	—	Harpreet S
2.	lubricating	ok	—	Harpreet S
3.	Adjusting	ok	—	Harpreet S
4.	Inspecting	ok	—	Harpreet S
5.	Any oil leakage	ok	—	Harpreet S
6.	Check filter element of hydraulic system	ok	—	Harpreet S
7.	Any creaks in the hoses	No	—	Harpreet S

Verified by Manager

Suitable actions taken on the grinding machine

The grinding machine was being used to grind chemicals. Excessive strain on the router caused the bearing seizure and the wear of the bearing sheet on which the bearing was attached, as well as bending the shaft as seen in the Figure 12.33. The bearing and shaft's life expectancy improved after implementing appropriate maintenance practices. The main cause of this defect was faulty belts, which damaged the bearing sheet and shaft. Corrective maintenance has been implemented to overcome the faults through the shaft replacement (Figure 12.34). More reliable and better quality belts have been used after this failure.

Following the installation of maintenance practices, belts have been monitored at regular intervals and replaced as necessary. As a result, the number of breakdowns has dropped.

Figure 12.33 Breakdown of the bearing sheet and shaft.

Figure 12.34 Shaft replacement.

Proper actions implemented to the grinding machine's rotary valve

The grinding machine's air lock or rotary valve, which regulates pressure in the chamber, was disabled because the nut had slipped out of the pin Figure 12.35. This occurred as a result of insufficient maintenance. The shaft was replaced (Figure 12.36) and the life of the rotary valve improved when the predictive maintenance was implemented, and it never failed during production.

Applicable maintenance measures implemented for the shaft

Figure 12.37 depicts a damaged shaft caused by weight limitations being exceeded. Motor belt failure, which caused the shaft's rotor to break down, was another cause of shaft failure. The shaft's rotor broke due to a lack of sufficient maintenance, and production was terminated. After determining the cause of failure, corrective maintenance was conducted, with higher-quality belts replacing lower-quality belts, and there were also regular inspections. Additionally, shaft life and production increased after maintenance practices were implemented.

Figure 12.35 Breakdown of the rotary valve.

Figure 12.36 Rotary valve replacement.

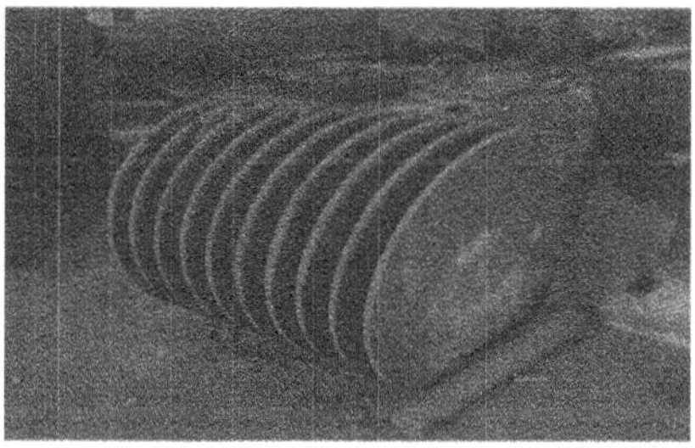

Figure 12.37 Breakdown of the shaft.

12.8 Appropriate Measures Implemented for the Hydraulic Bending Machine

Sheets were being rolled and bent in a hydraulic bending machine; and machine inspection was not being performed, so the hydraulic bending machine's seals wore out over time. Oil seals wore down due to lack of maintenance, resulting in significant oil leakage, causing the machine to break down and stop (Figure 12.38). A lack of or poor maintenance caused the leakage. Oil leaking harmed the performance of hydraulic bending and sharing machines. As a result of the leaking, the cylinder body and piston deteriorated. Through implementing preventive maintenance practices, seals were regularly evaluated and replaced (Figure 12.39) as needed, resulting in higher machine productivity and availability for production without downtime.

Figure 12.38 Leakage in the hydraulic bending machine.

Figure 12.39 Seal replacement.

As seen in the rolling machine, oil leakage from the gearbox resulted in accelerated gear tooth wear out, causing the machine to stop. The leakages resulted from a lack of maintenance (Figure 12.40) as the old oil seals were not replaced. The gear oil leaked due to worn oil seals, and the gear tooth broke due to a lack of lubricating oil. The oil seals were effectively monitored once maintenance practices were implemented, which increased machine life.

Figure 12.40 Leakage of gear oil in the rolling machine.

Wear of the cutter plate in the grinding machine

The wear of the cutter plate liner resulted in low machine efficiency and production, as indicated in the chemical grinding machine in Figures 12.41–12.44. The timely replacement of cutter plate and body liners increased the machine's output.

Figure 12.41 Wear out of the cutter plate.

Figure 12.42 Grinding machine with the new cutter plate.

Figure 12.43 Blower body of the grinding machine.

Figure 12.44 Wear out of the body liners in the grinding machine.

Table 12.6 Cost saved by replacing the body liners in a timely manner.

Time	Cost of the blower is INR 25,000 Cost of the liners is INR 4,000	Money saved (INR)
Jan. to June	25, 000 – 4,000	21,000
July to Dec.	25, 000 – 4,000	21,000
	Total money saved	42,000

The blower body was worn out due to a lack of timely replacement of body liners. The body liners were replaced on time once maintenance measures were implemented to extend the blower's life.

Cost saving: The proper maintenance program implementation in the plant resulted in significant cost savings. If body liners were replaced on time, the body of the blower could be saved. The cost of body liners was INR 4,000, and the cost of the blower body was INR 25,000 as shown in Table 12.6.

12.9 Results and Discussion

The SWOT analysis assisted in identifying organizational strengths and focusing on them to increase plant output. The analysis also helped identify the organization's weaknesses, thus, assisting in the reduction of organizational breakdowns, and provided opportunities to enhance the organization's equipment performance to accomplish goals. In the face of threats posed by competitors, the organization's maintenance taskforce got the impetus to stay on top of the market and adopt good maintenance practices to remain competitive.

As indicated in Table 12.7 and Figure 12.45, the number of breakdowns in the organization dropped significantly through the implementation of maintenance practices. The reduction in the number of breakdowns facilitated the achievement of enhanced availability of machines.

After implementing maintenance practices, the number of breakdowns was reduced significantly as indicated in Figure 12.45. As the breakdowns decreased, the manufacturing machines' availability increased as depicted in Table 12.8 and Figure 12.46. The increase in the availability of machines contributed to the enhancement in OEE of the plant.

With the application of strategic maintenance practices, the OEE increased from 61% to 83%, as shown in Table 12.9 and Figure 12.47. When machines performed without any failures, they increased plant production and facilitated the delivery of items on schedule to the market. They also helped improve product quality. The enhancement of OEE also aided in raising employee morale to carry out their responsibilities properly and achieve the most outstanding results. OEE improvement facilitated enhanced productivity of the plant.

Table 12.7 Month-wise breakdowns.

Month	Number of breakdowns
01-Jan	45
01-Feb	43
01-March	42
01-April	36
01-May	26
01-June	28
01-July	19
01-Aug	15
01-Sept	12
01-Oct	10
01-Nov	12
01-Dec	8

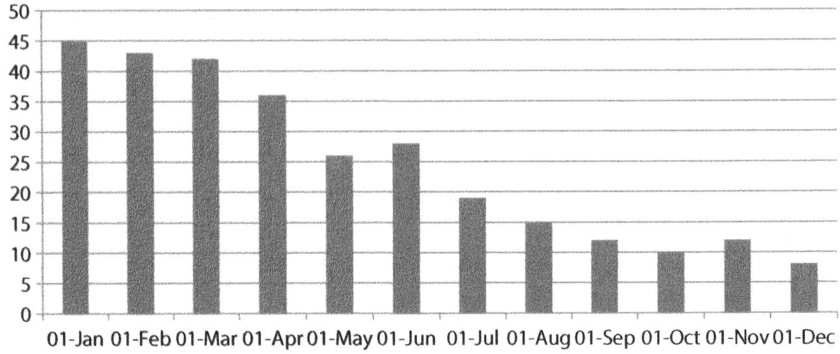

Figure 12.45 Reduction in breakdowns.

Table 12.8 Month-wise percentage of availability of the machines.

Month	In financial year-1 before the implementation of maintenance practices	In financial year-2 after the implementation of maintenance practices
January	61	78
February	65	85
March	60	88
April	70	80
May	68	82
June	50	90
July	84	92
August	62	95
September	65	91
October	68	90
November	63	94
December	66	95

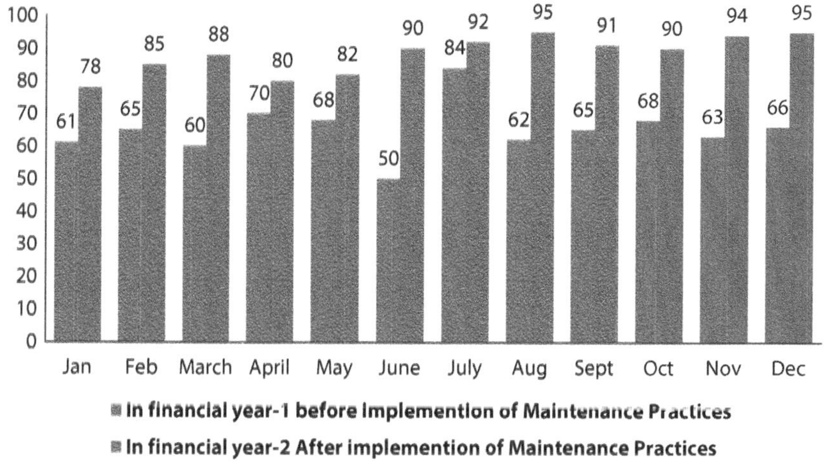

Figure 12.46 Month-wise percentage of availability of the machines.

Table 12.9 Month-wise OEE.

Month	Percentage of OEE
01-Jan	61
01-Feb	64
01-March	60
01-April	66
01-May	71
01-June	69
01-July	74
01-Aug	79
01-Sept	80
01-Oct	78
01-Nov	82
01-Dec	83

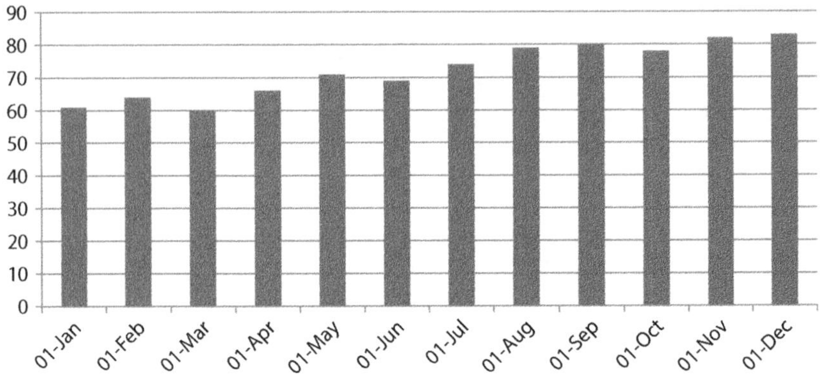

Figure 12.47 Improvements in OEE.

The reduction in plant breakdowns contributed significantly toward increasing the availability of machines and enhancing plant OEE through implementing maintenance practices. The product deliveries showed marked improvement, from 71% to 97%, as shown in Table 12.10 and Figure 12.48.

Table 12.10 Timely delivery of the product.

Month	Delivery compliance
01-Jan	71
01-Feb	73
01-March	78
01-April	81
01-May	82
01-June	85
01-July	87
01-Aug	88
01-Sept	90
01-Oct	92
01-Nov	95
01-Dec	97

Figure 12.48 On time delivery of the product increased.

12.10 Conclusions

The first contribution of this study was to identify areas where organizational performance needed improvement. The company's implementation of maintenance improvement practices revealed various issues that had potential for improvement to increase productivity, quality, availability, and reliability, among other company performance metrics. The company's phase-wise implementation of maintenance practices is the second contribution to the present work. In phase 1, the efforts directed toward improving performance parameters were evident. To reduce breakdowns, different preventive measures have been taken. The implementation of maintenance improvement practices has contributed to an increase in OEE and a decrease in MTTR. As a result of identifying predictive maintenance (Pd.M) issues and corrective maintenance (CM) issues in the implementation of maintenance practices, the manufacturing setup was improved, resulting in dramatic improvements in MTBF and a reduction in breakdowns. Apart from that, the importance of OEE in maintenance practices is acknowledged in enhancing organizational performance.

In phase 3 of maintenance practices implementation, importance has been given to increasing the machines' life and reducing the equipment's operating cost. By implementing all the above methods, the company reaped the maximum benefits by reducing the number of breakdowns. In the third phase, MTBF and OEE increased significantly through implementation of strategic maintenance practices.

In the second case study, SWOT analysis was implemented to determine the strengths, weaknesses, opportunities, and threats. As part of the methodology, the following steps were taken: identify the critical areas that influence production and breakdowns; examine the current situation to determine possible improvements; set up maintenance practices; compare and contrast old and new circumstances if there are positive changes; maintain the new scenario; if the situation does not improve, re-implement the appropriate maintenance practices. After implementing maintenance practices, the number of breakdowns was reduced significantly, and the availability of machines increased from 66% to 95%. Further, the delivery compliance increased from 71% to 88%, leading to improved performance of the organization.

References

Acharya, J.P. and Bhatt, M.J., A brief literature review on Total Productive Maintenance. *Int. J. Adv. Eng. Res. Dev.*, 1, 5, 1–8, 2014.

Adel, R., Smart services' quality scale. *Int. J. Prod. Qual. Manage.*, 35, 4, 429–446, 2022.

Ahuja, I.P.S. and Khamba, J.S., Total Productive Maintenance: Literature Review and Directions. *Int. J. Qual. Reliab. Manage.*, 25, 7, 709–756, 2008.

Ananth, G. and Vinayagam, B., Effectiveness improvement through total productive maintenance using particle swarm optimisation model for small and micro manufacturing enterprises. *Int. J. Prod. Qual. Manage.*, 16, 4, 473–503, 2015.

Attri, R., Khan, N.Z., Siddiquee, A.N., Khan, Z.A., Analysing the barriers to successful implementation of total quality management in Indian manufacturing organisations using best-worst method. *Int. J. Bus. Excellence*, 24, 2, 275–294, 2021.

Ben-Daya, M. and Duffuaa, S., Maintenance and quality: the missing link. *J. Qual. Maint. Eng.*, 1, 1, 20–26, 1995.

Benhayoun, L., Ayala, N.F., Le Dain, M.-A., SMEs innovating in collaborative networks: how does absorptive capacity matter for innovation performance in times of good partnership quality? *J. Manuf. Technol. Manage.*, 32, 8, 1578–1598, 2021, https://doi.org/10.1108/JMTM-11-2020-0439.

Bhadu, J., Bhamu, J., Singh, D., Sangwan, K.S., An ISM approach for lean implementation barriers in labour intensive Indian ceramic SMEs. *Int. J. Prod. Qual. Manage.*, 35, 3, 281–307, 2022.

Chandegra, P.N. and Deshpande, V.A., Cycle Time Reduction of PCF Gearbox using Total Productive Maintenance (TPM) Approach. *Int. J. Adv. Ind. Eng.*, 3, 2, 60–67, 2015.

Chandegra, P.N. and Deshpande, V.A., Cycle Time Reduction of PCF Gearbox using Total Productive Maintenance (TPM) Approach. *Int. J. Adv. Ind. Eng.*, 3, 2, 60–67, 2015.

Chen, J., Liu, L., Wang, Y., Business model innovation and growth of manufacturing SMEs: a social exchange perspective. *J. Manuf. Technol. Manage.*, 32, 2, 290–312, 2021, https://doi.org/10.1108/JMTM-03-2020-0089.

Chinese, D. and Ghirardo, G., Maintenance management in Italian manufacturing firms matters of size and matters of strategy. *J. Qual. Maint. Eng.*, 16, 2, 156–180, 2010.

Dixit, S. and Raj, T., Quantifying the influence of FMS on the productivity of a firm. *Int. J. Process Manage. Benchmarking*, 9, 4, 447–468, 2019.

Ghadimi, P., O'Neill, S., Wang, C., Sutherland, J.W., Analysis of enablers on the successful implementation of green manufacturing for Irish SMEs. *J. Manuf. Technol. Manage.*, 32, 1, 85–109, 2021, https://doi.org/10.1108/JMTM-10-2019-0382.

Hariharan, S.V., Benchmarking currency risk management practices of small and medium enterprises. *Int. J. Bus. Excellence*, 24, 3, 340–359, 2021.

Jain, A., Bhatti, R., Singh, H., Total productive maintenance (TPM): a proposed model for Indian SMEs. *Int. J. Mech. Prod. Eng. Res. Dev.*, 4, 1, 1–22, 2014a.

Jain, A., Bhatti, R., Singh, H., Sharma, S.K., Improvement of Indian SMEs through TPM Implementation – An Empirical Study, in: *Proceedings of International Conference on Advances in Mechanical Engineering AETAME-2013, Fourth International Joint Conference (Elsevier, ISTE, ACEEE, AMAE, ACEE and SEARCH) on Advances in Engineering and Technology*, pp. 786–792, 2013 December.

Jin, X., Siegel, D., Weiss, B.A., Gamel, E., Wang, W., Lee, J., Ni, J., The present status and future growth of maintenance in US manufacturing: results from a pilot survey. *Manuf. Rev.*, 3, Article no 10, 1–10, 2016, doi: 10.1051/mfreview/2016005.

Jutla, D., Bodorik, P., Dhaliwal, J., Supporting the e-business readiness of small and medium-sized enterprises: approaches and metrics. *Internet Res.*, 12, 2, 139–164, 2002, https://doi.org/10.1108/10662240210422512.

Kedar, A.P., Borikar, V.N., Kedar, A.P., Borikar, V.N., Critical success factors for effective implementation of TQM and TPM. *Int. J. Innov. Res. Sci. Technol.*, 2, 9, 160–164, 2016.

Kedaria, V.D. and Deshpande, V.A., Implementation of total productive maintenance methodology: a review. *Int. J. Emerg. Technol. Adv. Eng.*, 3, 4, 644–648, 2014.

Khtatbeh, M.M., Mahomed, A.S.B., Rahman, S.B.A., Mohamed, R., The role of equity in the relationship between job evaluation and employee performance in Jordan Industrial Estates. *Int. J. Prod. Qual. Manage.*, 35, 4, 576–598, 2022.

Kumar, R., Singh, K., Jain, S.K., Agility enhancement through agile manufacturing implementation: a case study. *TQM J.*, 34, 6, 1527–1546, 2021, https://doi.org/10.1108/TQM-05-2021-0147.

Kumar, S. and Rao, P., Financing patterns of SMEs in India during 2006 to 2013– an empirical analysis. *J. Small Bus. Entrepreneurship*, 28, 2, 97–131, 2016.

Kureshi, N., Qureshi, F., Sajid, A., Current health of quality management practices in service sector SME: A case study of Pakistan. *TQM J.*, 22, 3, 317–329, 2010.

Löfving, M., Melander, A., Elgh, F., Andersson, D., Implementing Hoshin Kanri in small manufacturing companies. *J. Manuf. Technol. Manage.*, 32, 9, 304–322, Emerald Publishing Limited, 2021, 1741-038X DOI 10.1108/JMTM-08-2020-0313.

Lundgren, C., Bokrantz, J., Skoogh, A., A strategy development process for Smart Maintenance implementation. *J. Manuf. Technol. Manage.*, 32, 9, 142–166, Emerald Publishing Limited, 2021. 1741-038X DOI 10.1108/JMTM-06-2020-0222.

Maelah, R. and Yadzid, N.H.N., Budgetary control, corporate culture and performance of small and medium enterprises (SMEs) in Malaysia. *Int. J. Globalisation Small Bus.*, 10, 1, 77–99, 2018.

Maletic, D., Maletic, M., Al-Najjar, B., Gomiscek, B., The role of maintenance regarding improving product quality and company's profitability: a case study. *2nd IFAC Workshop on Advanced Maintenance Engineering, Services and Technology*, Universidad de Sevilla, Sevilla, 22-23 November, pp. 7–12, 2012.

Moubray, J., 21st century maintenance organization part I: the asset management model. *Maint. Technol.*, 16, 2, 25–32, 2003.

Ng, K.C., Goh, G.G.G., Eze, U.C., Barriers in total productive maintenance implementation in a semiconductor manufacturing firm: A case study, in: *2012 IEEE International Conference on Industrial Engineering and Engineering Management*, pp. 377–381, IEEE, 2012.

Odoom, R., Narteh, B., Rand, J., Branding and outcomes in small and medium-sized enterprises (SMEs): a resource-capability approach. *J. Small Bus. Entrepreneurship*, 29, 3, 175–192, 2017.

Payan, J.M., Padín, C., Ferro, C., Svensson, G., Action and social alignment components of collaboration in SME business relationships. *J. Small Bus. Entrepreneurship*, 31, 6, 463–481, 2019.

Qamruzzaman, M. and Jianguo, W., SME financing innovation and SME development in Bangladesh: An application of ARDL. *J. Small Bus. Entrepreneurship*, 31, 6, 521–545, 2019.

Ramesh, N. and Ravi, A., Enhancing the performance of micro, small and medium sized cluster organisation through lean implementation. *Int. J. Prod. Qual. Manage.*, 21, 3, 325–342, 2017.

Sidhu, S.S., Singh, K., Ahuja, I.S., Ranking of implementation dimensions for maintenance practices in Northern Indian SMEs using integrated AHP-TOPSIS approach. *J. Small Bus. Entrepreneurship*, 34, 2, 175–194, 2022c. doi.org/10.1080/08276331.2020.1809220.

Sidhu, S.S., Singh, K., Ahuja, I.P.S., An empirical investigation of maintenance practices for enhancing manufacturing performance in small and medium enterprises of northern India. *J. Sci. Technol. Policy Manage.*, 13, 1, 132–153, 2022a.

Sidhu, S.S., Singh, K., Ahuja, I.S., Design of an intelligent decision-making system for maintenance practices using fuzzy inference system in northern Indian SMEs. *J. Qual. Maint. Eng.*, 28, 1, 154–179, 2022b, https://doi.org/10.1108/JQME-05-2020-0043.

Singh, G. and Ahuja, I.P.S., Selection of performance measure in JIT through fuzzy logic-based simulation. *Asian J. Manage. Sci. Appl.*, 1, 4, 388–402, 2014.

Singh, K. and Ahuja, I.S., Implementing TQM and TPM paradigms in Indian context: critical success factors and barriers. *Int. J. Technol. Policy Manage.*, 13, 3, 226–244, 2013.

Singh, P. and Kaur, C., Factors determining financial constraint of SMEs: a study of unorganized manufacturing enterprises in India. *J. Small Bus. Entrepreneurship*, 33, 3, 269–287, 2019.

Singh, R.K., Garg, S.K., Deshmukh, S.G., Implementation of information technology: evidences from Indian SMEs. *Int. J. Enterp. Network Manage.*, 2, 3, 248–67, 2008.

Singh, R.K., Garg, S.K., Deshmukh, S.G., The competitiveness of SMEs in a globalized economy. *Manage. Res. Rev.*, 33, 1, 54–65, 2010.

Srivastava, N., and Mondal, S., Maintenance practices in Indian manufacturing companies. *IJCSMS Int. J. Comput. Sci. Manage. Stud.*, 13, 5, 32–35, 2013.

Stentoft, J., Wickstrøm, K.A., Haug, A., Philipsen, K., Cost-driven motives to relocate manufacturing abroad among small- and medium-sized manufacturers: The influence of Industry 4.0. *J. Manuf. Technol. Manage.*, 32, 3, 646–666, 2021, https://doi.org/10.1108/JMTM-07-2019-0283.

Wagire, A.A., Rathore, A.P.S., Jain, R., Identification and prioritisation of challenges to Industry 4.0 adoption in the Indian manufacturing industry. *Int. J. Bus. Excellence*, 24, 2, 248–274, 2021.

Wang, C.N., Nguyen, V.T., Thai, H.T.N., Tran, N.N., Tran, T.L.A., Sustainable supplier selection process in edible oil production by a hybrid fuzzy analytical hierarchy process and green data envelopment analysis for the SMEs food processing industry. *Mathematics*, 6, 12, 1–16, 2018, https://doi.org/10.3390/math6120302.

Index

Also of Interest

Check out these related titles from Scrivener Publishing

Power Devices and Internet of Things for Intelligent System Design, Edited by Angsuman Sarkar and Arpan Deyasi, ISBN: 9781394305780. Unlock the potential of cutting-edge advancements in power electronics and IoT with "Power Devices and Internet of Things for Intelligent System Design," a vital resource that bridges the gap between industry and academia, inspiring innovative solutions across diverse fields like agriculture, healthcare, and security.

Intelligent Green Technologies for Smart Cities, Edited by Suman Lata Tripathi, Souvik Ganguli, Abhishek Kumar, and Tengiz Magradze, ISBN: 9781119816065. Presenting the concepts and fundamentals of smart cities and developing "green" technologies, this volume, written and edited by a global team of experts, also goes into the practical applications that can be utilized across multiple disciplines and industries, for both the engineer and the student.

Hybrid Intelligent Optimization Approaches for Smart Energy: A Practical Approach, Edited by Senthilkumar Mohan, A. John, Sanjeevikumar Padmanaban, and Yasir Hamid, ISBN: 9781119821243. Written and edited by a group of experts in the field, this is a comprehensive and up-to-date description of current energy optimization techniques, such as artificial intelligence techniques, machine learning, deep learning, and IoT techniques and their future trends.

NANODEVICES FOR INTEGRATED CIRCUIT DESIGN, Edited by Suman Lata Tripathi, Abhishek Kumar, K. Srinivasa Rao, and Prasantha R. Mudimela, ISBN: 9781394185788. Written and edited by a team of experts in the field, this important new volume broadly covers the design of nano-devices and their integrated applications in digital and analog integrated circuits (IC) design.

EXPLAINABLE MACHINE LEARNING MODELS AND ARCHITECTURES: Real-Time System Implementation, Edited by Suman Lata Tripathi and Mufti Mahmud, ISBN: 9781394185849. This cutting-edge new volume covers the hardware architecture implementation, the software implementation approach, and the efficient hardware of machine learning applications.

MACHINE LEARNING TECHNIQUES FOR VLSI CHIP DESIGN, Edited by Abhishek Kumar, Suman Lata Tripathi, and K. Srinivasa Rao, ISBN: 9781119910398. This cutting-edge new volume covers the hardware architecture implementation, the software implementation approach, and the efficient hardware of machine learning applications with FPGA or CMOS circuits, and many other aspects and applications of machine learning techniques for VLSI chip design.

INTELLIGENT GREEN TECHNOLOGIES FOR SMART CITIES, Edited by Suman Lata Tripathi, Souvik Ganguli, Abhishek Kumar, and Tengiz Magradze, ISBN: 9781119816065. Presenting the concepts and fundamentals of smart cities and developing "green" technologies, this volume, written and edited by a global team of experts, also goes into the practical applications that can be utilized across multiple disciplines and industries, for both the engineer and the student.

DESIGN AND DEVELOPMENT OF EFFICIENT ENERGY SYSTEMS, Edited by Suman Lata Tripathi, Dushyant Kumar Singh, Sanjeevikumar Padmanaban, and P. Raja, ISBN: 9781119761631. Covering the concepts and fundamentals of efficient energy systems, this volume, written and edited by a global team of experts, also goes into the practical applications that can be utilized across multiple industries, for both the engineer and the student.

Electrical and Electronic Devices, Circuits, and Materials: Technical Challenges and Solutions, Edited by Suman Lata Tripathi, Parvej Ahmad Alvi, and Umashankar Subramaniam, ISBN: 9781119750369. Covering every aspect of the design and improvement needed for solid-state electronic devices and circuit and their reliability issues, this new volume also includes overall system design for all kinds of analog and digital applications and developments in power systems.

MODELING AND OPTIMIZATION OF OPTICAL COMMUNICATION NETWORKS, Edited by Chandra Singh, Rathishchandra R Gatti, K.V.S.S.S.S. Sairam, and Ashish Singh, ISBN: 9781119839200. *Modeling and Optimization of Optical Communication Networks* is a comprehensive and authoritative book that delves into the optical networks, principles, technologies, and practical applications of optical networks, equipping readers with the knowledge needed to design, implement, and optimize optical networks for various applications, from telecommunications to scientific research. With its comprehensive coverage and up-to-date insights, this book serves as an essential reference in the field of optical networks.

WIRELESS COMMUNICATION SECURITY: Mobile and Network Security Protocols, Edited by Manju Khari, Manisha Bharti, and M. Niranjanamurthy, ISBN: 9781119777144. Presenting the concepts and advances of wireless communication security, this volume, written and edited by a global team of experts, also goes into the practical applications for the engineer, student, and other industry professionals.

ADVANCES IN DATA SCIENCE AND ANALYTICS, Edited by M. Niranjanamurthy, Hemant Kumar Gianey, and Amir H. Gandomi, ISBN: 9781119791881. Presenting the concepts and advances of data science and analytics, this volume, written and edited by a global team of experts, also goes into the practical applications that can be utilized across multiple disciplines and industries, for both the engineer and the student, focusing on machining learning, big data, business intelligence, and analytics.

ARTIFICIAL INTELLIGENCE AND DATA MINING IN SECURITY FRAMEWORKS, Edited by Neeraj Bhargava, Ritu Bhargava, Pramod Singh Rathore, and Rashmi Agrawal, ISBN: 9781119760405. Written and edited by a team of experts in the field, this outstanding new volume offers solutions to the problems of security, outlining the concepts behind allowing computers to learn from experience and understand the world in terms of a hierarchy of concepts.

MACHINE LEARNING AND DATA SCIENCE: Fundamentals and Applications, Edited by Prateek Agrawal, Charu Gupta, Anand Sharma, Vishu Madaan, and Nisheeth Joshi, ISBN: 9781119775614. Written and edited by a team of experts in the field, this collection of papers reflects the most up-to-date and comprehensive current state of machine learning and data science for industry, government, and academia.

MEDICAL IMAGING, Edited by H. S. Sanjay, and M. Niranjanamurthy, ISBN: 9781119785392. Written and edited by a team of experts in the field, this is the most comprehensive and up-to-date study of and reference for the practical applications of medical imaging for engineers, scientists, students, and medical professionals.

SECURITY ISSUES AND PRIVACY CONCERNS IN INDUSTRY 4.0 APPLICATIONS, Edited by Shibin David, R. S. Anand, V. Jeyakrishnan, and M. Niranjanamurthy, ISBN: 9781119775621. Written and edited by a team of international experts, this is the most comprehensive and up-to-date coverage of the security and privacy issues surrounding Industry 4.0 applications, a must-have for any library.

CYBER SECURITY AND DIGITAL FORENSICS: Challenges and Future Trends, Edited by Mangesh M. Ghonge, Sabyasachi Pramanik, Ramchandra Mangrulkar, and Dac-Nhuong Le, ISBN: 9781119795636. Written and edited by a team of world renowned experts in the field, this groundbreaking new volume covers key technical topics and gives readers a comprehensive understanding of the latest research findings in cyber security and digital forensics.

DEEP LEARNING APPROACHES TO CLOUD SECURITY, Edited by Pramod Singh Rathore, Vishal Dutt, Rashmi Agrawal, Satya Murthy Sasubilli, and Srinivasa Rao Swarna, ISBN: 9781119760528. Covering one of the most important subjects to our society today, this editorial team delves into solutions taken from evolving deep learning approaches, solutions allow computers to learn from experience and understand the world in terms of a hierarchy of concepts.

MACHINE LEARNING TECHNIQUES AND ANALYTICS FOR CLOUD SECURITY, Edited by Rajdeep Chakraborty, Anupam Ghosh and Jyotsna Kumar Mandal, ISBN: 9781119762256. This book covers new methods, surveys, case studies, and policy with almost all machine learning techniques and analytics for cloud security solutions.

SECURITY DESIGNS FOR THE CLOUD, IOT AND SOCIAL NETWORKING, Edited by Dac-Nhuong Le, Chintin Bhatt and Mani Madhukar, ISBN: 9781119592266. The book provides cutting-edge research that delivers insights into the tools, opportunities, novel strategies, techniques, and challenges for handling security issues in cloud computing, Internet of Things and social networking.

DESIGN AND ANALYSIS OF SECURITY PROTOCOLS FOR COMMUNICATION, Edited by Dinesh Goyal, S. Balamurugan, Sheng-Lung Peng and O.P. Verma, ISBN: 9781119555643. The book combines analysis and comparison of various security protocols such as HTTP, SMTP, RTP, RTCP, FTP, UDP for mobile or multimedia streaming security protocol.

SMART GRIDS FOR SMART CITIES VOLUME 2: Real-Time Applications in Smart Cities, Edited by O.V. Gnana Swathika, K. Karthikeyan, and Sanjeevikumar Padmanaban, ISBN: 9781394215874. Written and edited by a team of experts in the field, this second volume in a two-volume set focuses on an interdisciplinary perspective on the financial, environmental, and other benefits of smart grid technologies and solutions for smart cities.

SMART GRIDS AND INTERNET OF THINGS, Edited by Sanjeevikumar Padmanaban, Jens Bo Holm-Nielsen, Rajesh Kumar Dhanaraj, Malathy Sathyamoorthy, and Balamurugan Balusamy, ISBN: 9781119812449. Written and edited by a team of international professionals, this groundbreaking new volume covers the latest technologies in automation, tracking, energy distribution and consumption of Internet of Things (IoT) devices with smart grids.

Printed and bound by CPI Group (UK) Ltd, Croydon, CR0 4YY
17/09/2026
14964276-0001